AF688439

Dyes from Discard: Upcycling Tea Waste for Sustainable Fabric Dyeing

Edited by

Vinitha Moolchand Thadhani
Ceylon Ayurveda Pvt. Ltd.
Havelock Road
Colombo, Sri Lanka

Dyes from Discard: Upcycling Tea Waste for Sustainable Fabric Dyeing

Editor: Vinitha Moolchand Thadhani

ISBN (Online): 979-8-89881-540-0

ISBN (Print): 979-8-89881-541-7

ISBN (Paperback): 979-8-89881-542-4

Published by Bentham Science Publishers Pte. Ltd. Singapore, in collaboration with Eureka Conferences, USA.

First published in 2026.

Bentham Science Publishers Pte. Ltd.
No. 9 Raffles Place
Office No. 26-01
Singapore 048619
Singapore
Email: subscriptions@benthamscience.net

CONTENTS

FOREWORD

Over the past few decades, the textile industry has stood at a crossroads. On one side lies the convenience and low cost of synthetic dyes; on the other, the urgent need to address their environmental and health burdens. Textile dyeing is widely recognized as one of the major contributors to global water pollution, and yet synthetic dyes still dominate commercial practice. Natural dyes, despite their rich heritage, represent only a small share of the market and are often perceived as difficult to scale, inconsistent in shade, or less durable in performance.

The reasons for this imbalance are well known to both researchers and industry professionals: challenges in sustainable sourcing, cost, limited availability, and, crucially, issues of reproducibility and fastness. These very challenges, however, have sparked some of the most creative and impactful research in sustainable colorants. This book, focused on the valorization of tea waste as a fabric dye, is an excellent example of such innovation in action.

Tea is among the most widely consumed beverages worldwide, and tea processing generates large quantities of by-products and waste. While tea (*Camellia sinensis*) has been explored as a natural dye in academic studies, much of this work has remained at the laboratory scale, with limited translation to industrial application. The authors of this volume address precisely this gap. They demonstrate how waste fractions from Ready-to-Drink (RTD) tea manufacturing can be transformed into a robust, scalable, and commercially viable natural dyeing system.

One of the strengths of this book is its clear and logical structure. The opening chapters situate the work within a broader context, tracing the history and rise of synthetic dyes and outlining their impacts on human health and the environment. This is complemented by an accessible overview of contemporary textile dyeing technologies. Together, these chapters provide the reader with a solid foundation for understanding why alternative dyeing approaches are urgently needed.

The book then guides the reader from why to how. It presents a balanced discussion of natural versus synthetic dyes, outlining their respective advantages and limitations, and exploring the role of bio-based textile materials and green chemistry strategies in the fashion and textile sectors. This framing reinforces the idea that dye innovation cannot be isolated from wider efforts to build circular, low-impact textile systems.

At the heart of the volume lies its central contribution: the use of tea waste for sustainable textile coloration. The authors describe in detail how black, oolong, white, and green tea waste streams can be converted into stable dye extracts, explaining the optimization of extraction conditions, application methods, and fastness properties. Importantly, the work does not stop at bench-scale experiments. The narrative follows the process from laboratory studies to pilot trials and then to bulk-scale implementation, offering a rare and valuable roadmap for practitioners interested in moving from concept to practice.

Equally compelling is the way this book embodies the principles of circular economy and industrial symbiosis. By viewing discarded tea fractions from RTD manufacturing as a resource rather than a waste, the authors demonstrate how one industry's by-product can become another's raw material. The fact that the approach has already attracted interest and implementation from reputed brands underscores both its technical soundness and its commercial relevance.

This volume will be of great interest to a wide audience. Researchers and students will appreciate its scientific rigor and clear explanation of process parameters. Textile and fashion professionals will find practical guidance on integrating natural dyes into existing production systems without compromising performance. Sustainability practitioners and circular economy specialists will recognize in this work a concrete example of how waste valorization can be successfully embedded in real industrial contexts.

Ultimately, this book is more than a technical guide; it is a testament to what can be achieved when thoughtful science, meticulous experimentation, and a genuine commitment to sustainability come together. By addressing long-standing barriers such as reproducibility, consistency, and scalability, the authors bring natural dyeing one step closer to mainstream industrial reality.

I am pleased to commend this book to readers and to support its contribution to the advancement of sustainable textile coloration and circular resource use.

Luqman Jameel Rather
College of Sericulture, Textile
and Biomass Science, Southwest University
Chongqing, China

PREFACE

Although textile dyeing is the second-largest polluter of water globally, the market for natural dyes in textiles is growing, driven by increasing consumer demand, environmental and health concerns, and increasingly stringent government regulations associated with synthetic dyes.

Still, the market share of natural dyes is only 5% of that of synthetic dyes. The annual global consumption of synthetic dyes in the textile industry exceeded 650,000 metric tons, while global natural dye consumption annually is only 32,000 metric tons.

Sustainable sourcing, limited availability, and cost, reproducibility of the natural dyeing process, and durability of natural dyes are the primary reasons why the synthetic dye markets still dominate.

Currently, natural indigo, turmeric, and madder are the most frequently used natural dyes in textiles, food, and cosmeceuticals industries. For example, the dry madder stems cost around Rs. 4-7/Kg, but they have only 5-10 % of the active ingredients (pure dye). Incorporating the extraction process cost, the effective cost of madder dye goes up to Rs. 40-140/Kg. Thus, it is challenging to replace the synthetic dyes, which mostly cost under two figures USD per Kg. Considering these factors, natural dyes from waste or industrial by-products are being focused on.

There are ample reports on tea (*Camellia sinensis*) as a natural fabric dye, and few reports on textile dyes from waste tea. However, most of these studies are specialized studies conducted on a laboratory scale, which may not be sufficient for industrial applications. Thus, tea dye has not paved its way to industrial application, even after being the most consumed beverage after water.

Herein, we discuss how tea dye can be obtained from upscaling waste products of the RTD tea manufacturing process, its chemical composition, and the method of application as an effective fabric dye. We also establish its required fastness properties as a fabric dye. Further methods of obtaining different colours using tea dye and minute amounts of synthetic dyes are also discussed. The lab scale results were upscaled to the sample scale and then finally to the bulk scale in the textile industry.

The idea of using waste from one factory as a raw material for another is a win-win scenario, promoting resource efficiency and circular economy principles. The findings of this research have been implemented by various reputable brands worldwide.

The following is a chapter-by-chapter synopsis of the eight chapters of the book.

Chapter 1: History of Natural Dyes and Their Recent Advances

This chapter examines the history of natural dyes from antiquity to the present. The chapter also discusses how contemporary science has improved the natural dye extraction process and its wide application, and how natural dyes are now a vital element of science, sustainability, and creativity.

Chapter 2: Sources of Natural Dyes

This chapter provides a comprehensive overview of natural dyes, demonstrating how their diverse sources, structures, and applications make them valuable for textiles and a wide range of other applications.

Chapter 3: Natural dyes vs Synthetic dyes: Pros, Cons, and Sustainability

This chapter explains how both types of dyes work, how to use them, and how their color strength, consistency, fastness, and ease of use differ. It also highlights the environmental impact and carbon footprint of both types of dyes. Understanding their benefits and drawbacks can assist us in selecting dyeing methods that are more sustainable and efficient.

Chapter 4: Bio-based textile materials towards Green Chemistry in the Fashion Industry

This chapter highlights how more eco-friendly manufacturing procedures can be developed utilizing green chemistry, bioengineering, new dyeing, and recycling approaches for cost-effective production and for limiting greenhouse gas emissions.

Chapter 5: Valorization of Tea Waste for Sustainable Fabric Dye

This chapter presents a comprehensive study of the innovative process of transforming the tea waste of black, oolong, white, and green leaves into a sustainable fabric dye. Furthermore, the parameters for optimizing the tea extraction process are discussed.

Chapter 6: Method of Application of Waste Tea Dye

The chapter presents a refined approach that integrates the functional properties of tea with traditional dyeing techniques, covering substrate pretreatment, different dyeing methods, and post-treatment.

Chapter 7: A success story of Transforming Ready-to-Drink Tea waste into fabric dye

This study establishes the discarded phenolic fraction of the Ready to Drink (RTD) tea manufacturing process as a fabric dye with the required fastness properties, offering insights for researchers, circular engineers, and industries, and presenting a sustainable alternative method for using waste tea as a dye in the textile sector.

Further, the main challenges of commercial application of natural dyes, namely, supply chain, cost, reproducibility, and consistency, have been overcome by utilizing the waste of the RTD manufacturing plant.

This book is the culmination of all the know-how of using tea waste as a sustainable fabric dye. We hope it will help our readers immensely by disseminating scientific knowledge to them step by step.

We wish to sincerely thank Bentham Publications for providing a platform for the dissemination of the information published in this book to a broad audience. We are also very thankful to every author for their essential contributions that have made our journey worthwhile.

Vinitha Moolchand Thadhani
Ceylon Ayurveda Pvt. Ltd.
Havelock Road, Colombo
Sri Lanka

List of Contributors

Belal Uddin Rabbi	Department of Wet Process Engineering, Bangladesh University of Textiles (BUTEX), Dhaka, Bangladesh
Gehan Amaratunga	Department of Engineering, Electrical Engineering Division, University of Cambridge, Cambridge, UK
Mahdieh Dehghan	Department of Textile Engineering, Yazd University, Yazd, Iran
Mohammad Khajeh Mehrizi	Department of Textile Engineering, Yazd University, Yazd, Iran
Muhammad Usama	Department of Biochemistry, Government College University Faisalabad, Faisalabad, Pakistan
Naheed Akhtar	H.E.J. Research Institute of Chemistry, International Center for Chemical and Biological Sciences, University of Karachi, Karachi, Pakistan
Shahid Adeel	Department of Applied Chemistry, Government College University Faisalabad, Faisalabad, Pakistan
Sultana Bedoura	Department of Dyes and Chemical Engineering, Bangladesh University of Textiles (BUTEX), Dhaka, Bangladesh
Syed G. Musharraf	H.E.J. Research Institute of Chemistry, International Center for Chemical and Biological Sciences, University of Karachi, Karachi, Pakistan Dr. Panjwani Center for Molecular Medicine and Drug Research, International Center for Chemical and Biological Sciences, University of Karachi, Karachi, Pakistan School of Chemistry and Pharmaceutical Sciences, Guangxi Normal University, Guilin, Guangxi, China
Vinitha M. Thadhani	Ceylon Ayurveda Pvt. Ltd, Havelock Road, Colombo, Sri Lanka
Zahra Shahi	Department of Textile Engineering, Yazd University, Yazd, Iran

CHAPTER 1

History of Natural Dyes and Their Recent Advances

Naheed Akhtar[1,*], **Syed G. Musharraf**[1,2,3] and **Vinitha M. Thadhani**[4]

[1] *H.E.J. Research Institute of Chemistry, International Center for Chemical and Biological Sciences, University of Karachi, Karachi, Pakistan*

[2] *Dr. Panjwani Center for Molecular Medicine and Drug Research, International Center for Chemical and Biological Sciences, University of Karachi, Karachi, Pakistan*

[3] *School of Chemistry and Pharmaceutical Sciences, Guangxi Normal University, Guilin, Guangxi, China*

[4] *Ceylon Ayurveda Pvt. Ltd, Havelock Road, Colombo, Sri Lanka*

Abstract: This chapter explores the historical journey of natural dyes from ancient times to modern times. The use of natural dyes dates back to ancient societies, especially those in the Mediterranean, North Africa, Nigeria, South America, and Europe. Colorful fabrics for clothing, rituals, and commerce were created by people in these areas using plants, minerals, and insects. Various methods for extracting dyes and extending the life of colors were developed over time, utilizing compounds known as mordants. The chapter also discusses how contemporary science has utilized these dyes to enhance the process of extracting and utilizing them. Techniques such as microwave-assisted and supercritical fluid extraction enable us to extract more dye in less time and with fewer chemicals. These innovative methods are safer for both humans and the environment. Natural dyes are now utilized in various industries, including textiles, pharmaceuticals, solar technology, and food production. They find applications in cancer therapies, UV-protective materials, sensors, and solar cells. This chapter demonstrates how natural dyes, with their long and rich history, are now a vital element of science, sustainability, and creativity.

Keywords: Cochineal, Deodorizing finishing, Mordants, Moth resistant, Natural dyes.

INTRODUCTION

Natural dyes are colorants derived from natural ingredients, including leaves, roots, bark, flowers, fruits, minerals, and even insects. Long before the development of synthetic dyes, people around the globe utilized natural dyes to

* **Corresponding author Naheed Akhtar:** H.E.J. Research Institute of Chemistry, International Center for Chemical and Biological Sciences, University of Karachi, Karachi, Pakistan; E-mail: naheedakhtar41@yahoo.com

Vinitha Moolchand Thadhani (Ed.)

color their clothing, paint items, and represent their culture and beliefs. Natural dyeing was a traditional art practiced by generations in North Africa, Nigeria, South America, and Europe.

Historically, manufacturing and utilizing these dyes requires a significant amount of time and effort. To remove the color, they had to boil or soak the plants for several days. They employed mordants to keep the color on the fabric. These allowed the dye to adhere to the textile, increasing the color's brilliance and longevity. Mordants used in different regions ranged from metal salts to plant-based solutions.

As time passed, researchers began to seek more effective techniques to simplify and enhance the natural dyeing efficiency. They developed new extraction techniques, such as solid-phase microextraction and microwave-assisted extraction, which save time and reduce the use of hazardous chemicals. These approaches also make the dyeing process more environmentally friendly.

Today, natural colors are utilized for more than simply garments. They are being utilized in various exciting new applications, including natural colorants in food, medical imaging, antibacterial materials, and solar panels. This chapter discusses the rich history, scientific advancements, and many present applications of natural dyes.

HISTORY OF NATURAL DYES

Natural dyes have had a significant impact on shaping and rewriting history around the world. They bear an important historical witness that broadens the horizon, allowing us to look back at the wonderful past and forward to an optimistic future [1]. For thousands of years, natural dyes have been a popular choice for coloring textiles, food, and various products due to their eco-friendliness, accessibility, cost-effectiveness, and safety [2].

From the dawn of humanity, nature has consistently prevailed over synthetic materials, as those were the sole options available. Now that naturally created materials outperform synthetics, they are prioritized. Color has consistently influenced the evolution of diverse human cultures across the globe. It influences all aspects of our lives, greatly shaping the clothing we choose and the furnishings in our homes. In the past, artists have traditionally used natural dyes derived from minerals, mollusks, insects, and plants in their works [3]. A brief history of natural dyes is depicted in Fig. (**1**).

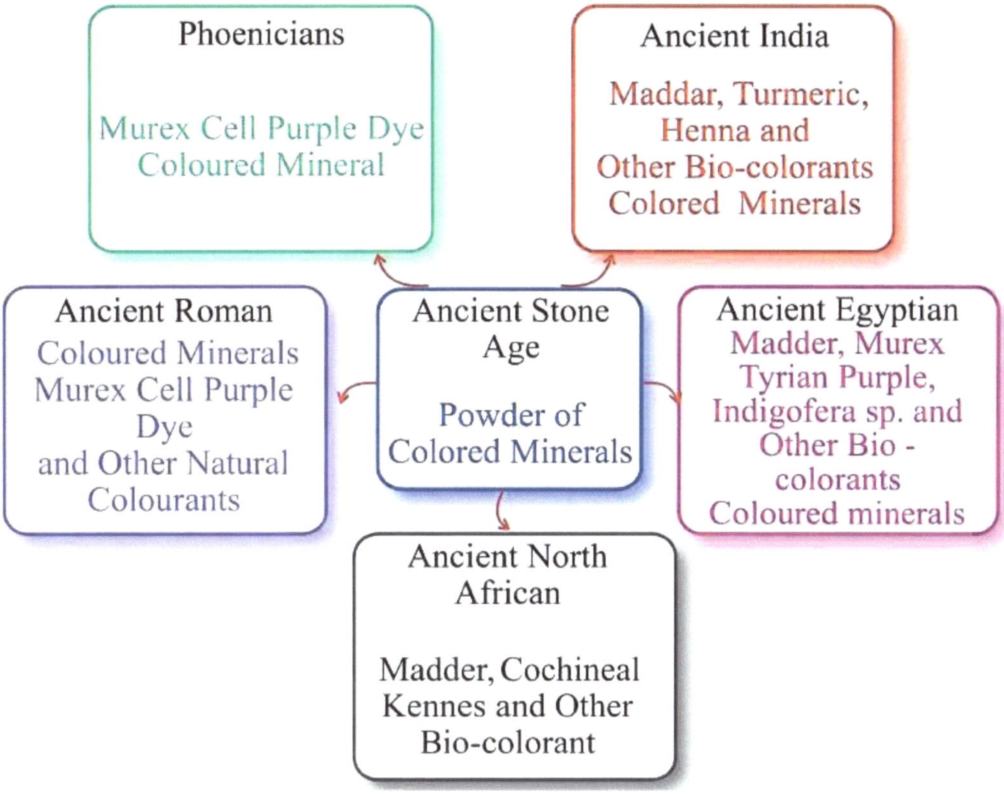

Fig. (1). A brief history of natural dyes.

History of Natural Dyes in the Ancient Mediterranean World

Ancient Mediterranean iconography gives a wealth of information about the intricate and colorful world of textiles. The 'Ancient Mediterranean', a broad term referring to a period from the Bronze Age to the late Roman Empire (1500 BCE-500 CE), indicates a time of substantial cultural and political interaction between Southern Europe, Western Asia, and North Africa [4]. Plant roots, insects, and sea snails provided a wide range of natural hues to ancient Mediterranean civilizations. Expertise and understanding are required for dye extraction and application. The colorants were used as both dyes and dye pigments, with the majority being dye lakes. Important groups of ancient dyes were anthraquinone reds (*Kermes vermilio*, *Rubia tinctorum*), redwoods (*Caesalpinia sappan*), and flavylium/anthocyanin reds (*e.g.*, *Dracaena draco*); indigo (*e.g.*, *Indigofera tinctoria*) and anthocyanins (*Commelina communis*) for blue; Tyrian purple (*e.g.*, *Purpura*, *Murex brandaris*), orchil (*e.g.*, *Roccella* sp.), and folium purple (*Chrozophora tinctoria*); and flavonoid (*e.g.*, *Reseda luteola*), carotenoid (*Crocus sativus*), and chalcone/aurone (*Coreopsis* sp.) yellows [5].

In North Africa

Africa boasts a vibrant cultural heritage in the art of dyeing. West African nations like Ghana have a rich tradition of textile dyeing, harnessing natural hues from plants and minerals. Kente fabric from Ghana exemplifies the rich heritage of traditional African dyeing methods. The centuries-old practice of dyeing wool with natural dyes is especially popular in Moroccan places like Fez. The Egyptians dyed fabrics for clothing and ceremonial purposes with natural resources such as plants, minerals, and insects. Early civilizations utilized madder root and indigo as some of the first natural sources for obtaining red and blue dyes [2].

The bio-colorants, such as madder (*Rubia tinctoria*), cochineal (*Dactylopius coccus*), and kermes (*Kermes vermilio*), were used by ancient North Africans to produce dyes and pigments. However, these were far more economical and were frequently utilized for dyeing, medieval miniature painting, and cosmetics. The Egyptians were aware of these because they excelled at weaving, as many inscriptions praise the gods' garments and bandages for the deceased, which were primarily dyed with archil, a purple color derived from specific marine algae found on Mediterranean Sea rocks. Alkanet, a red dye derived from the roots of *Alkanna tinctoria* and *Rubia tinctorum*. Woad (*Isatis tinctoria*) and indigo (*Indigofera* species) are the blue dyes that are derived from the leaves of these plants through a fermentation process [3, 6].

History Ofddyes in Nigeria

Nigeria has an extensive and diverse dye history that represents the country's cultural heritage as well as its trading connections with neighboring countries and distant lands. It also demonstrates how external factors, such as colonialism and globalization, have affected the Nigerian dyeing industry. Plant-based dyes have a long history in Nigeria, which is linked to the country's rich cultural legacy and diverse flora. For ages, Nigeria's traditional communities have used dyes for fabrics, crafts, and ceremonies, which are obtained from plant sources. In addition to meeting basic demands, this strategy significantly facilitated the expression of artistic traditions and cultural identity. Nigeria's biological and geographic diversity has produced a wide variety of plant species that can be used to extract dyes. Many ethnic groups in Nigeria have been known to employ plant-based dyes, each with its own plants and dyeing methods. Indigo, derived from the *Indigofera* plant, is one of Nigeria's most widely used plant-based colors. Nigeria has a long tradition of indigo dyeing, and the Yoruba people are especially adept at it. Indigo plants are fermented to extract pigment for the dyeing process, which produces hues that range from deep blue to green [7]. In Nigeria, the *Baphia nitida* tree, commonly referred to as "Camwood" or "Osun," is another noteworthy natural source of dye [8]. Several Nigerian civilizations, particularly the Yoruba and Igbo, have utilized the dye produced by this tree to create textiles with red and pink hues. Camwood dye has cultural value and is widely employed in rituals and traditional practices. The bark, roots, and leaves of *Lannea welwitschii* have been used to make colors in Nigeria. Boiling, fermenting, or soaking are common methods used to extract dyes. Notably, the usage of these plant pigments was not only aesthetically pleasing but also conveyed social status within groups and had symbolic importance. Numerous variables have influenced the evolution of plant-based dyes in Nigeria. Traditional plant-based dyes, which provided more color variation, consistency, and application efficiency, were less often used during the colonial era due to the arrival of synthetic dyes. Traditional expertise and methods for plant-based dyeing consequently started to wane over time. On the other hand, eco-friendly and traditional dyeing procedures have grown in popularity in recent years. Nigerian artisanal, artistic, and cultural preservationists are attempting to bring back the usage of plant-based dyes for aesthetic, cultural, and environmental grounds. The establishment and sustainable harvesting of dye plants is another way that this revival boosts local economies [2].

In South America

The Americas also have a rich history of dyeing. The Incas and Aztecs, for example, used dyes derived from cochineal insects and annatto seeds to create vibrant shades of red, orange, and yellow. Textiles, paintings, ceramics, and body ornamentation were among the many uses for these dyes. Within communities, early dyeing techniques were occasionally well-kept secrets that were passed down through the years. Knowledge of dyeing methods spread as civilizations engaged in trade and conquest, leading to innovations and cross-cultural interactions. Rich dyeing traditions were found throughout the Americas. Cloth dyeing has a long history in Peru's Andes, with practices dating back to pre-Columbian times among indigenous tribes in Central and South America. Vibrant fabrics are still created with natural dyes obtained from plants and insects. Furthermore, a variety of natural dyes originating from regional plants and animals are utilized in Oaxaca and Chiapas, which are examples of Mexican textile traditions., utilizing methods that have been passed down over the years [2].

In Europe

Scotland has a long history of coloring wool using natural plant dyes, such as lichen, moss, and heather. The country is well-known for its traditional tartans. In current Scottish dyeing techniques, synthetic dyes are also utilized to achieve a greater variety of hues. Turkish carpet weaving uses vibrant hues created through natural dying methods. These were typically obtained from locally sourced substances such as insect dye and madder root. In Mesopotamia, the Sumerians and Babylonians used turmeric, saffron, and other plant extracts to produce a range of colors. In various ancient societies, such as the Phoenicians and Romans, the famed “Royal Purple” dye, created from the secretions of the mollusk Murex, was highly esteemed and associated with royalty [2, 9].

Dyeing Practices in Other Parts of the World

Many societies throughout history have utilized natural resources to produce exciting and enduring dyes. These habits, which were rooted in necessity and tradition, paved the way for the development of dyeing techniques. India is well-known across Asia for its widespread usage of natural colors. The country employs plants such as madder root, turmeric, and indigo to produce a variety of vivid colors. Moreover, Rajasthan and Gujarat are home to traditional techniques, including block printing and tie-dying. Furthermore, complicated resist dyeing techniques are employed in Japanese dyeing techniques, such as Shibori, to create gorgeous textile patterns. Indigo dyeing holds tremendous cultural value, particularly in Tokushima Prefecture. Ancient Chinese culture had a significant influence on the development of early dyeing technologies. To create a diverse range of hues, the Chinese created methods for obtaining dyes from plants, including indigo, woad, and gardenia. They developed expertise in the skill of silk dyeing, which was popularized by the Silk Road trading networks [2].

Drastic Conditions Used in Natural Dyes

Extraction Methods

Various techniques are used to extract natural colorants from natural sources. Following the drying and grinding of the plant material, a suitable solvent is chosen in accordance with the target dye and extracted using both traditional and modern methods. Chromatographic or other separation techniques are employed to isolate the targeted chemicals used in dyeing, and spectroscopic methods are utilized to characterize the isolated compounds. Modern extraction technologies are superior to traditional methods because they yield higher quantities at a lower cost. Fig. (**2**) demonstrates different techniques for extracting natural dyes. Important Contemporary techniques are discussed briefly as follows.

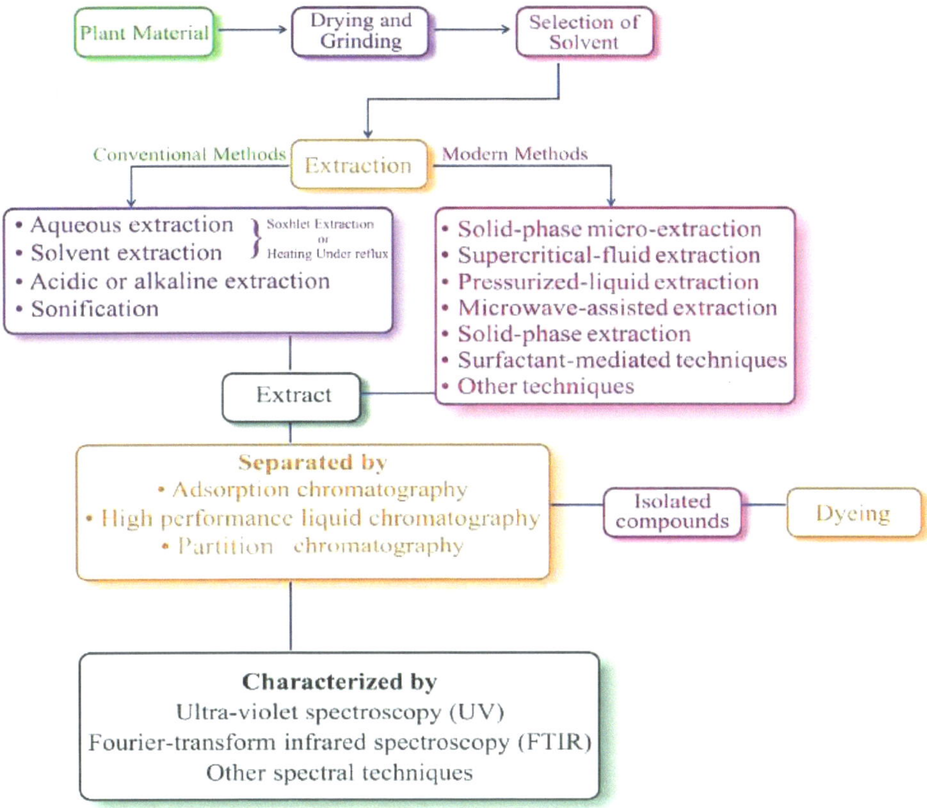

Fig. (2). Extraction methods of natural dyes.

The initial stage of extraction consists of gathering, drying, and grinding the plant material in order to produce a uniform mixture and maximize its surface area for interaction with the solvent. Selection of an appropriate solvent for the extraction or isolation of plant material is the next most important step. Polar solvents, such as methanol, ethanol, or ethyl acetate, can be used to extract hydrophilic compounds. In contrast, dichloromethane or a 1:1 combination of dichloromethane and methanol can be used for lipophilic chemical extraction. Depending on the polarity and thermal stability of the target compound, various procedures such as sonication, reflux heating, Soxhlet extraction, and others are often employed. Some modern extraction methods, such as solid-phase microextraction, supercritical-fluid extraction, pressurized-liquid extraction, microwave-assisted extraction, solid-phase extraction, and surfactant-mediated techniques, are also employed due to their high yields and ease of extract collection. Extraction yields mixtures of substances, which are then separated using techniques such as adsorption chromatography, Thin-Layer Chromatography (TLC), and High-Performance Liquid Chromatography (HPLC). The compounds are then analyzed using spectroscopic techniques such as ultraviolet (UV) spectroscopy and Fourier-transform infrared (FTIR) spectroscopy, among others.

Nonetheless, a great deal of research has already been done on the process of extracting colourants from plants. The anthocyanin content is absorbed by passing the extract through an ion exchange column. It is subsequently eluted with dimethyl formamide (DMF), acetone, or an alkali. Furthermore, a patent issued in 1998 describes a process using pressurized gases like propane and/or butane to extract carotenoid pigments from natural starting materials that have already been dried, and then adding organic entraining agents to help finish the extraction process. This technique produces extremely concentrated carotenoid pigments in high yield. Extraction of tea dye from the Broken Mixed Fanning (BMF), usually a tea leftover, is described in a patent in 2021. The BMF is extracted into hot water and treated with hydrolyzing enzymes, such as Viscozyme. After several steps, the hydrophobic, comparatively non-polar groups, such as flavones/isoflavones glycosides, were extracted, and the resulting powder is used as the tea dye, which has been successfully commercialized [10, 11]. Mantell *et al.* investigated the extraction of anthocyanin pigments from red grape pomace under high pressures using carbon dioxide and other solvents such as methanol or water. The colorimetric approach was used for quantification after a number of extraction parameters, such as temperature, pressure, solvent flow rate, co-solvent percentage, solvent type, and extraction time, were examined for the best results. The optimal settings for the highest yield were 20 mol% methanol, 100 bar pressure, 60 °C temperature, and 22 mmol/min flow rate. Anthocyanin dyes were produced from red pomace by Bechtold *et al.* and used to dye textiles in distilled water. With a material to liquor ratio (M: L) of 1:20 for 60 minutes at 95 °C. The extraction of raw dye material of pomegranate peel was done using a 1:5 material-to-solvent ratio (ethanol-water) for 60 minutes at 60 °C. The resulting filtrate was distilled for 3 hours at 70 °C in a Soxhlet system, yielding a concentrated dye suitable for dyeing. The dye (a mixture of gallic acid, ellagic acid, quercetin, and rutin compounds) was extracted from fresh eucalyptus leaves, dried in sunlight for one month, and crumbled with a blender using the reflux technique; 70 g of crumbled eucalyptus leaves were mixed in a liter of distilled water and refluxed for one hour. The evaporation of filtration was done under low pressure, and the resulting filtrate was dried and utilized for the coloration of wool and silk. Tannins were extracted from the aqueous extract of the tea plant by mixing 100 ml of water with 2 and 5 grams of tea powder, stirring, heating, bringing to a boil for 30 minutes, allowing it to stand for 15 minutes, and then filtering and using the solution to dye cotton. Anthraquinone dyes were extracted from *Cassia tora* L. seed using various pH buffer solutions (pH 2-11) for three days at room temperature in a material-to-liquor ratio of 1:10. The *Cassia tora* L. extract solution (natural dye solution) prepared at pH 9 was found to have the maximum K/S ratio and a yellowish red solution for coloring cotton and silk. Additionally, the yellow pigments were extracted from coffee sludges through aqueous extraction for 90 minutes at 90 °C, and the resulting mixture was assessed for color fastness and deodorizing qualities [3].

Modern Methods of Extraction

Modern extraction techniques are far better than traditional, outdated methods because they provide high outcomes at low cost. The following is a quick discussion of important modern approaches.

Solid-Phase Micro Extraction

Sample input, extraction, concentration, and sampling are all combined into a single procedure in solvent-free extraction. This approach consists of two steps: (a) separating the t compound of interest from the sample and extraction media, and (b) storing the desorbed, concentrated extract in the instrument. This process requires less solvent and produces enhanced components. It is quite simple. The dearth of commercially accessible stationary phases is one of the biggest drawbacks of solid phase micro extraction.

Supercritical Fluid Extraction

The method involves the solubility of the dye of interest in supercritical fluid, which is usually CO_2, C_2H_4, C_3H_8, or C_4H_{10}. In a high-pressure extraction container, the dye component comes into contact with the supercritical fluid. It is a sophisticated separation technology that relies on the critical temperature and solvating capacity of gases. CO_2 is the most preferred gas for supercritical fluid extraction because it is less toxic, has a lower critical temperature, is less flammable, and is readily available in its pure form at the cheapest cost. CO_2 is used at temperatures ranging from 40 to 800 °C and pressures ranging from 35 to 75 MPa. The advantage of this approach is that minor changes in pressure and temperature result in extremely high variations in solvent density and solubility. However, this technology is not cost-effective, and CO_2 is a suitable alternative due to its low cost in many applications.

Pressurized Liquid Extraction

This approach uses higher temperatures and moderate to high pressures to remove color components more effectively. This method saves time and uses less solvent. This extraction begins with the sample being dispersed with an inert substance or sand, and the sample is then poured into the vessel. An appropriate solvent is added to the jar in a static step, 0.5-21 min, along with heating, and dynamic extraction will be performed with the flow of solvent. The efficiency of this process is determined by solvents, temperature, pressure, and the composition of the adsorbents used. The main advantage of this technique is that it allows for fast extraction with a small amount of solvent; however, it also requires expensive equipment for cleanup.

Microwave-assisted Extraction

Extraction is performed using several solvents and microwave power levels at 264, 400, and 600 W. The microwave rations are used to extract the color pigments by heating the sample and the desired solvent for the extraction. The desired component is introduced in the solvent through the sample. Thermally unstable components can also be extracted using this technique due to its rapidness. This extraction method depends on the nature of the time, strength of microwave radiation, temperature, sample, components to be extracted, and nature of the solvent, *etc*. This approach is employed in two configurations: closed vessel system and open vessel system. Thermally unstable compounds can not escape from the vessel system, and less use of solvent is required, which is a key benefit of the closed vessel system. But this approach is limited to the selected number of samples. An open vessel system operates at atmospheric pressure, allowing reagents to be added during processing. The open vessel method operates securely at atmospheric pressure, and reagents can be added during processing. However, open vessel systems cannot process a large number of samples simultaneously, whereas closed systems can process many samples at once.

Solid Phase Extraction

In this method, an adsorbent is selected on the basis of the desired dye. Both liquid-liquid extraction and solid phase extraction are comparable to each other because of the material distribution between the two phases. This approach uses a variety of adsorbents, including silica-based, carbon-based, and clay-based resins. The purification and concentration of components can be enhanced with a solid adsorbent. This approach offers several advantages over liquid-liquid extraction, including higher recoveries, reduced emulsion formation, and simplified operation.

Surfactant-Mediated Techniques

In this approach, nonionic surfactants are used to remove and concentrate hydrophobic components from water. The aqueous medium helps the constituent to enter the surfactant media. This strategy is also applied in combination with microwave or ultrasonic technology. The use of an exact element with a higher affinity for the surfactant medium is the key advantage of this method [12].

Mordanting and Dyeing Methods

Following extraction processing, natural colorants are applied to textiles, with or without the use of mordants. The use of natural colourants in textile functionalization and colouring is becoming more popular., from their early usage in conventional textile dyeing to novel and advanced procedures that have recently become popular. Mordents are chemicals essential for most natural dyes to adhere more effectively to textiles. Fig. (**3**) shows the schematic representation of the mordanting and dyeing profile.

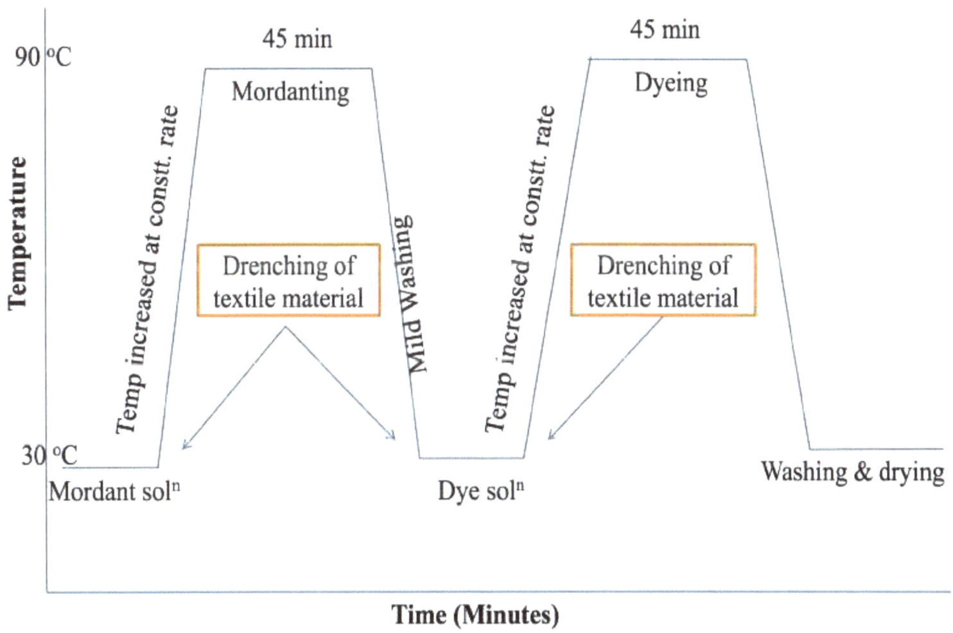

Fig. (3). Schematic representation of the mordanting and dyeing profile.

Mordanting Methods

To increase the substantivity of natural colorants in textiles, various metal salts or other chemicals or compounds, known as mordants, are used in conjunction with the colorants. Depending on when the mordents are applied, mordanting is divided into three categories: pre-, post-, and simultaneous mordanting [12]. Some mordent and mordanting methods are shown in Fig. (**4**).

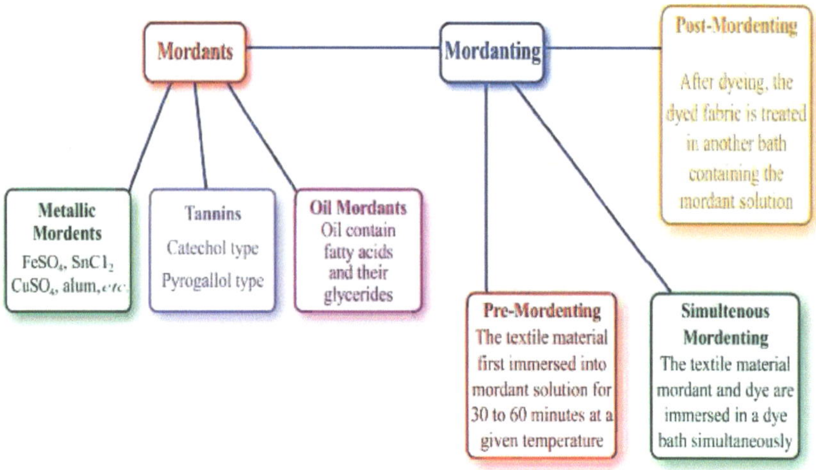

Fig. (4). Mordent and Mordanting Methods.

Pre-mordanting

Pre-mordanting refers to the treatment of textile materials with mordants before dyeing, which allows the mordants to bind to the textile substance exclusively and at specific areas. This method of processing natural colorants on textiles results in a proper layering of dye, mordant, and textile substance. A color is produced when the metal combines with the textile's surface and dye from opposing sides. That is lightfast, washable, and rubbing resistant. In this process, chelating complexation increases the light fastness of colored materials and guarantees that light photons in the complex dissipate their energy properly. Chelating complexation in this procedure ensures that the light photons dissipate their energy properly in a complex, which improves the fastness of light in colored substances. The use of optimal resources in pre-mordanting makes this more sustainable for the environment, flora, and animals.

Meta-mordanting/Simultaneous Mordanting

Both mordants and dyes are dissolved simultaneously in the dye bath for dyeing [13]. This type of processing wastes a significant amount of dye and mordant materials due to their complex chemical interactions. Mordants occupy some textile material sites, while others are directly occupied by dye compounds, resulting in irregularity of dyes. The complexation may occur between textile and dye, textile and mordant, or mordant and dyes in three different ways. In three different ways. Between textiles and mordants, between textiles and dyes, and between dyes and mordants, resulting in an excess of dye effluent entering the ecosystem and posing a threat to sustainability issues.

Post-mordanting

During this process, the bare textile material is first dyed or colored, and then a mordant is applied. By complexing mordants with dye molecules on textile materials' surfaces, this process primarily extends the shade spectrum. Maintaining color fastness may not be possible with this process.

Dyeing Methods

Conventional Dyeing System

Since its inception, cloth dyeing has been a traditional practice. High-temperature dye bath processing was applied directly to textiles. Recent decades have witnessed several advancements in the dyeing context, including the methods of printing, assessment of efficient mordants, and processes of dyeing. Numerous patents explained the application of indigo dye in textiles. In which the material is first pre-treated with a mordant, which is considered environmentally safe. After this, an inert atmosphere is employed for the application of indigo dye and then oxidizing the dye by saturating the surface with cold water. Disperse dyes from supercritical CO_2 can be used to dye cellulosic textiles by treating them with an auxiliary that facilitates dye absorption, usually polyethylene glycol. A technique for coloring textiles has been developed that utilizes chemically generated gels to create precise pattern prints and color designs with considerable flexibility. It may be utilized with standard dyeing and printing equipment. Dyeing also developed over time as dyeing conditions were optimized, and more recently, new methods such as enzyme processing and plasma treatment have emerged. While some textiles, such as cotton, require a mordant for improved dye binding, a dye solution is used to dye the wool and silk. Textile materials are processed using baths at high temperatures. Using chemically created gels, this textile coloring technique offers a high degree of flexibility in creating color and print patterns with standard printing and dyeing [12].

Advanced Dyeing Systems

Due to their superior results compared to traditional dyeing, advanced technologies and procedures have gained popularity in recent years. In the highly advanced textile sector, plasma treatment and ultrasonic dyeing are contemporary, cutting-edge, and environmentally friendly techniques. Ultrasound waves and plasma, sometimes referred to as the fourth state of matter, have enough energy to influence the energy of the dye bath's constituent parts. The results in ultrasound-assisted dyeing depend on the cavitation phenomena and some mechanical sound effects like dispersion or breaking of high molecular weight, diffusion or increase in the diffusion of dye in the fiber, degassing or ejection of air from fiber capillaries, and strong liquid agitation. Many workers may have noticed an acceleration in dyeing rates, which could be the result of all three elements working together.

Recent years have seen the adoption of sophisticated dyeing technologies and procedures, as they produce better outcomes than traditional methods. The scientifically advanced textile industry uses environmentally benign and technically superior techniques, including plasma treatment and ultrasonic dyeing. UV and gamma radiation treatments are also beneficial for color fastness [12].

Mechanism of Dyeing

Understanding adsorption isotherms, thermodynamics, and kinetics is crucial to comprehending the dyeing mechanism. Adsorption isotherms are the Temkin isotherm, the Hill isotherm model, the Redlich–Peterson isotherm model, the Freundlich adsorption isotherm model, and the Langmuir adsorption isotherm [3, 12]. Fig. (**5**) displays a dyeing method flow chart with several mordanting approaches.

Mordants Used in Natural Dyes

A mordant solution must be applied to the textile fibres in order to use the vast majority of natural dyes, which are mordant dyes (generally a metal salt). The fiber absorbs the solution, which enables the metal ion to combine with the proper functional groups in the fiber's structure. The dye and the mordant–fiber complex combine throughout the dyeing process to create an insoluble, vividly colored species. The mordant has a significant impact on the final color achieved and guarantees the dye's brightness and washability. Mordants include ions of copper, iron, tin, chromium, or aluminium. Several clubmoss species, including Diphasiastrum complanatum L. from Finland, *Diaphasiastrum alpinium* L., and *Huperzia selago* L. from Scotland, can provide aluminium. In contrast, iron can be found in bog bottom sediments or surface scum, as well as in minerals. These enhance the hues, shield the fibers, and/or facilitate the absorption of the mordants [14].

Metallic Salt (e.g., Aluminum, Zinc, Copper, Tin, Cobalt, Potassium)

Iron, copper, tin, aluminium, and chromium metal salts are employed. Alum, copper sulphate, ferrous sulphate, potassium dichromate, and stannous chloride are a few of the commonly used mordants. These metallic mordants are further separated into two categories, namely brightening mordants and dulling mordants, based on the final color generated with the natural dyes. The common examples of brightening mordants are white alum, tin, and potassium, whereas the examples of dulling mordants are ferrous and copper sulphates [15].

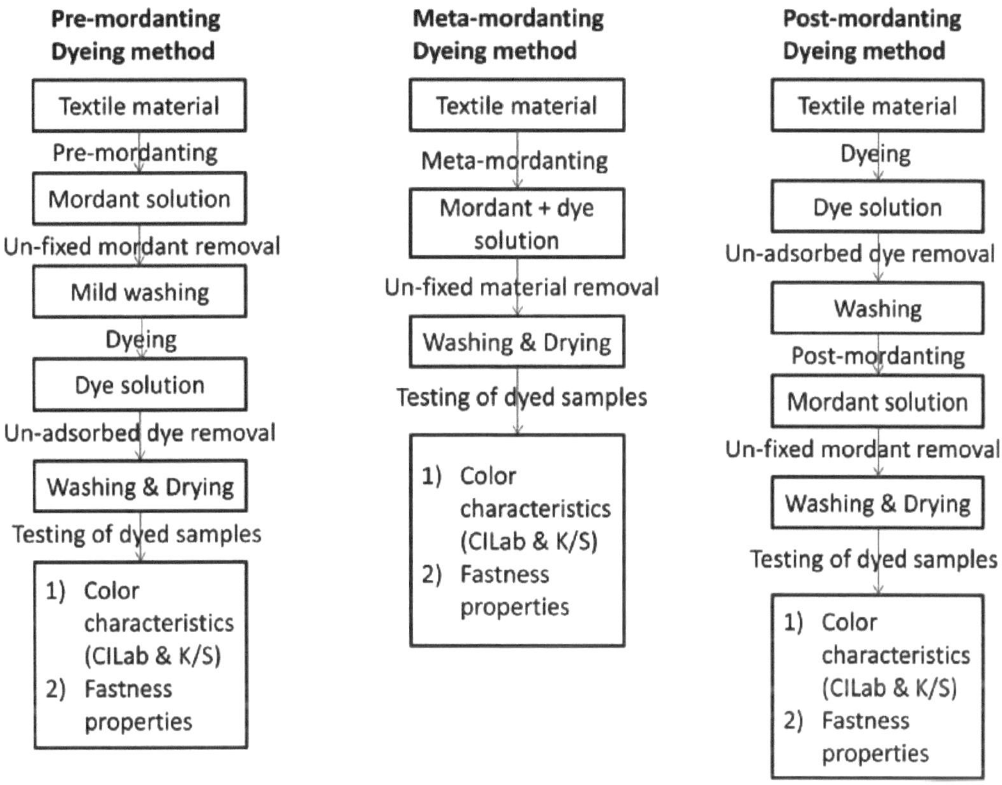

Fig. (5). Flow chart for dyeing methods with different mordanting techniques.

Bio-Mordant (e.g Madder, Myrobalan, Berries, Kermes, Cochineal, Tannic Acid)

Bio-mordants enhance the natural colors' ability to adhere to the cloth, resulting in fabrics with superior washing fastness. When using natural dyes to color textiles, bio-mordants are an excellent substitute for creating a more environmentally friendly method. Research into using natural bio-mordants instead of metal ions has gained popularity recently due to their eco-friendliness and effectiveness in enhancing the natural dyeing of textile materials.

Bio-mordants are plentiful, safe, and have no negative environmental effects. Bio-mordants come from a variety of natural sources, particularly plants. Due to their tannin content and chemical composition, high-tannin plants can be used as a mordant; however, they are also frequently categorised as significant dyes that don't need a metallic mordant. Pomegranate, turmeric, henna, acacia, arjun, gallnut, myrobalan, harmal, neem, and so forth are examples of bio-mordants. Several crucial factors, including the method, temperature, pH, treatment duration, and liquor ratio (M: L), significantly impact the mordanting process with bio-mordants. Furthermore, several crucial factors, including the dyeing process and dyeing parameters such as the M: L ratio, pH, temperature, and treatment duration, significantly impact bio-mordant-assisted natural dyeing [16].

Oil Mordant that Creates a Complex with Metallic Salts

The primary usage of oil mordants is in the madder dyeing of Turkey Red. With alum serving as the primary mordant, the primary function of oil mordants is complex. Alum is readily removed from the treated cloth because it dissolves in water and has no attraction for cotton. Fatty acids like palmitic acid are found in naturally occurring oils. When the COOH group of these acids reacts with metal ions, it transforms into COOM, where M stands for the metal. The treatment of oils with concentrated metal binding capability was later discovered to have a higher binding capacity than natural oil due to the reaction of sulfuric acid and metal, forming SO_3M. When combined with mordant dyes like madder, this bonded metal can create a complex that gives Turkey Red its exceptional hue and fastness. Consequently, Turkey Red Oils were the name given to the sulphonated oils (TRO). Nowadays, sulphonated castor oil is referred to as TRO. Additionally, TRO is a precursor to contemporary synthetic surfactants [15].

Recent Advances in Dying Textiles with Natural Dyes

There is a lot of space to investigate modern technology for natural textile dyeing by increasing the dyeing process's efficiency in terms of cost, time, and energy. Textile scientists and colorists attempted to add a physical or chemical treatment to textile dyeing in order to improve the color performance of naturally dyed fabrics. These unique therapies are often divided into two categories: dry treatment and moist treatment. Surface modification agents with cationic and anionic groups are examples of wet treatment used on cotton and wool textiles. In addition to being eco-friendly, the application of these agents significantly enhanced the color, colorfastness, and practical qualities of dyed textiles. Chitosan, a naturally occurring biopolymer with unique chemical and biological properties, has been used in several studies as a surface modification agent to enhance dyeing performance on cotton, silk, and wool fabrics by increasing binding sites and improving dye absorption. Additional research revealed that chitosan pretreatment enhanced the antibacterial and antioxidant properties of cotton and wool fibers.

Modern surface modification techniques are considered superior to conventional chemical modification procedures for the dry treatment of textiles. They are also associated with a more ecologically friendly textile processing methodology. Ozone, radiation, and plasma therapies are a few types of dry treatments. Fibers treated with Low-Temperature Plasma (LTP) exhibit improved surface properties, but their bulk properties remain unchanged. Through the cavitation effect, ultrasonic radiation can optimize the extraction and coloring process. Ozone can be utilised in a variety of chemical reactions and is a potent oxidizing agent. In general, it has also been studied as a bleaching method for altering fibers before dying. Fabric treated with gamma radiation may help fibers resist shrinking and wrinkling, increase the rate at which dye is absorbed, bind more dye to fibers, and improve the shade of the dyed fabric. This enhances colorant extraction and deepens the extract's color. This analysis did not include additional materials that are employed in conjunction with chitosan, such as various biopolymers, alginates, cyclodextrins, gums, and their derivatives. Overall, the introduction of these novel textile dyeing and finishing methods holds great promise for reducing the ecological impact of current processes without compromising eco-safety standards. Modern technology will undoubtedly open up more exciting opportunities for applications involving natural coloring [17].

Enzymatic processing has been employed as an environmentally responsible and sustainable technique for functionalizing and colouring textiles. To assess the impact of enzymatic treatment on colour features, cotton and silk were pretreated with tannic acid complexed with three enzymes: lipase, diastase, and protease-amylase. The cotton and silk were then dyed using natural colours. Cotton and silk textiles treated with enzymes showed faster colour adsorption kinetics and overall higher adsorption than samples that were not treated. The textile business is developing more quickly thanks to modern dyeing techniques and advanced technologies because they are sustainable and eco-friendly [3].

Additionally, natural colors were utilized in the food industry, cosmetics (catechu, henna), pharmaceuticals (rhubarb, saffron), and apparel industries (Annatto, Curcumin, and Cochineal) [3].

Humans currently use a greater variety of synthetic colorants for various practical purposes in modern culture. However, the environment and human health are at risk from the overuse of synthetic colors that resemble azo and benzidine and are not biodegradable and carcinogenic. Even if natural dyes do not have these negative impacts, they nevertheless need to be addressed because of their weak binding ability and low yield. Therefore, it is imperative that more studies be done and that natural resource output be optimized. From an Indian perspective, there are sufficient plants, minerals, and animals that produce dyes, and that can be produced in large quantities through cultivation or processing. It is possible to investigate using insects and bacteria to produce colorants. The performance of colorants can also be enhanced by investigating the usage of natural mordants. These days, natural dyes are used in every industry where synthetic dyes play a significant role. Some of these natural dyes have even replaced synthetic dyes due to their desired sustainability [12].

Fig. (**6**) illustrates the wide range of applications for natural dyes.

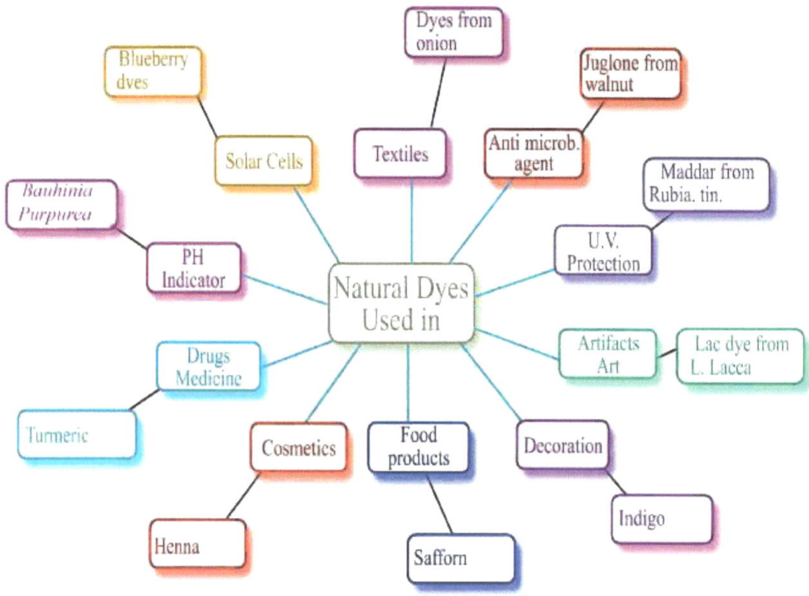

Fig. (6). Application of natural dyes in various fields.

Natural dyes play a significant part in food coloring, moth-resistant and insect-repellent fabrics and artefacts, deodorizing finishing, antimicrobial refined textiles, and UV shielding textiles. Apart from these applications, natural pigments and dyes are employed in the following fields [12]:

Cell Imaging

Cell imaging can also benefit from the use of natural fluorescent dyes. A luminous dye derived from beetroot serves as an example of this type of application [12].

Dye-sensitized Solar Cells

DSSC technology has advanced significantly with the use of natural dyes. Natural colours, including anthocyanins, carotene, flavonoids, and chlorophyll, are increasingly being extracted using simple methods, displacing costly metal complex sensitizers and conventional chemical synthesis methods. Natural dyes are, therefore, inexpensive, plentiful, and harmless materials that pose no threat to the environment [12, 18].

Indicator Substances

Currently, the food and other industries use a range of cutting-edge packaging materials composed of biodegradable, environmentally acceptable polymers, including polylactide and polyhydroxybutyrate, incorporating natural dyes or pigments. The color-changing property of natural colorants represents them as a strong indicator. In response to external influences, it contributes to the uniqueness of intelligent packaging; in this instance, color changes reveal the materials' longevity. Therefore, using natural colorants helps create new materials that are completely biodegradable [12, 19].

Biosensors

Natural dyes like indigo and others can be used as biosensors because of their electrochemical activity [12].

Cancer Therapy

Punica granatum, a natural color, showed promise in combating cancer. Dye-based nanomedicine and fluorescence optical imaging are also used to diagnose cancer [12].

Corrosion Inhibitors

Plant-derived pigments such as Lawson, curcumin, and berberine can effectively inhibit corrosion in a variety of metals across a range of conditions because of N, S, and O heteroatoms in their structure. For future viewpoints, these are crucial, reasonably priced green corrosion inhibitors [12, 20].

Antimicrobial Finished Textiles

For these microorganisms, every cloth offers a growth medium. Cotton and wool are examples of natural materials that are more vulnerable to microbial development and dust mite infestation because they retain oxygen, water, and nutrients. Clothes in a closet, curtains, carpets, bed linens, kitchen and bathroom towels, and even cushions and mattresses can harbour microorganisms. Human skin cells have several dust mites that can be in contact with different substances like cloth, towels, and sheets because a number of germs develop on human skin.

Similar to a home, a hospital has a vast number of textiles and the additional risk of significant microbial transmission.

To eradicate these microbes, antimicrobial drugs of both natural and synthetic origin were used. Natural-origin agents should be preferred in textile finishing due to their environmentally favorable qualities. In the past, fabrics were effectively colored and finished with antibacterial properties using natural dyes. Successful antimicrobial finishing of wool fibers was the outcome of an investigation into the impact of *Rheum emodi* L. as a dye and the activity of its dyed wool yarns against two bacterial (*Escherichia coli* and *Staphylococcus aureus*) and two fungal species (*Candida albicans* and *Candida tropicalis*). Using the micro-broth dilution method, disc diffusion assay, and growth curve studies, the antimicrobial activity of catechu in solution and the percentage microbial reduction of dyed wool samples against *Escherichia coli* MTCC 443, *Staphylococcus aureus* MTCC 902, *Candida albicans* ATCC 10261, and *Candida tropicalis* ATCC 750 were evaluated. On human erythrocytes, hemolytic activity was also investigated to rule out the likelihood of further associated cytotoxicity. Catechu's observed antibacterial properties and low cytotoxicity suggested that the dye is a suitable antimicrobial ingredient for creating biologically valuable textiles and apparel.

Microorganisms can proliferate due to the intrinsic qualities of textile fibers. Additionally, microorganisms may grow due to chemical reactions and the substrate's structure. Warm, humid conditions continue to exacerbate the issues. Microbe infestation results in pathogen cross-infection and odor production where the skin is in close proximity to the cloth. Scientists were able to experiment with a variety of natural and manmade materials with antibacterial finishes. The way society viewed health and safety changed in tandem with our understanding of functional finishes and synthetic fibers. Many people concentrated on educating themselves and safeguarding themselves from dangerous diseases as a result of this rise in health consciousness. Soon, protecting the wearer from bacteria will become more important for textiles with antibacterial finishes than merely shielding the garment from fiber deterioration.

For these microorganisms, every cloth offers a growth medium. Natural textiles, such as cotton and wool, are especially vulnerable to microbial development and dust mite invasion because they retain oxygen, water, and nutrients. Because of the constant stream of patients, especially those with infectious disorders, several experts have concentrated on creating finishes specifically for hospital use. Cross-transmission of illnesses and other health problems is a risk for both patients and staff. Most of these microbes are spread by different textiles from one person to another. Finding safe and long-lasting antimicrobial treatments is even more crucial now that drug-resistant germs are becoming more prevalent. Antimicrobial properties are inherent in several elements and natural substances. The antimicrobial textile market is dominated by metallic compounds and heavy metals. Cadmium, copper, silver, and mercury are all strong antimicrobials. Metal-based finishes can be used as a reusable finish because they are reasonably resistant to repeated washings. There are numerous non-metallic, natural, antibacterial coatings available. Chitosan is a naturally occurring antibacterial finish, obtained from the crustacean shells by deacetylation of chitin. Research has demonstrated that chitosan is effective against both Gram-positive and Gram-negative bacteria. In response to issues such as these, researchers have experimented with finishes currently available on the market. To mitigate the negative effects and enhance the finishes' antimicrobial efficiency, biologically active material is allowed to treat the fabric. The possibility of drug-resistant bacteria emerging from the finish is decreased by combining finishes. Adding antimicrobial chemicals to synthetic fibers while they are being spun is another trend in antimicrobial finish experimentation.

For many years, people have utilised natural dyes for their colouring and medicinal properties. However, only recently have their structures and protective qualities been acknowledged. It has recently been demonstrated that several antimicrobial dyes have been extracted from plants with unique medicinal values. Curcumin from turmeric, naphthoquinones like lawsone from *Lawsonia inermis*, juglone from walnut, lapachol from taigu, catechin from *Acacia catechu*, and several anthraquinones, such as those found in *Rubia tinctorum*, *Rubia cordifolia*, and *Rheum emodi*, are examples of common natural dyes that have demonstrated antimicrobial activity. Natural dyes from *Quercus infectoria* and *Punica granatum* are reported to possess potent antimicrobial properties due to their high content of bioactive phytochemicals.

Large volumes of hazardous chemicals are released into the environment as a result of the manufacture and usage of synthetic antibacterial agents, which are the root cause of numerous skin conditions and related illnesses. Antimicrobial textiles associated with non-toxic and environmentally acceptable biologically active chemicals are highly sought after in order to lessen the risks associated with the application of artificial antibacterial agents. Natural products can serve as an appealing and environmentally responsible alternative to synthetic medications in textile applications due to their comparatively lower incidence of adverse effects. Despite the existence of multiple natural antimicrobial agents, they can considerably lower the risk of infection, particularly when utilised in direct and indirect environments or in close contact with patients. Naturally occurring dyes have made a wonderful contribution to environmentally friendly textile finishing because of their remarkable antimicrobial properties [21].

UV Protective Textiles

Ultraviolet radiation, which constitutes a small portion of the solar spectrum, affects the metabolism of all living organisms. Exposure to this UV radiation might result in skin cancer, and tanning—clothing and creams with sunscreen offer defense against the damaging effects of UV rays. Modifications to the fabric's production parameters can be used to create UV-protective textiles by using the appropriate absorber of light and methods for finishing [22].

Three natural yellow dyes—curcumin, Rheum emodi, and Gardenia yellow were excellently employed for the dyeing and functionalization of silk, supplying UV-resistant fabrics. Under various settings, dye derived from eucalyptus leaves was applied to wool fabric using two padding procedures, the pad-batch and pad-dry processes. It was found that as the dye concentration increased, the range of wool fabric's UVPF value was very good to outstanding. The primary constituent of the aqueous extract of honeysuckle is chlorogenic acid, which was examined for its UV-protective qualities on wool. Applying honeysuckle extract to wool finishing demonstrated excellent ultraviolet transmittance in both the UV-A and UV-B bands, indicating that it might be used as a naturally occurring UV absorber. Cotton textiles were treated with natural plant colorants, madder (*Rubia tinctorum*) and indigo (*Indigofera tinctoria*), as well as the natural insect colorant cochineal (*Dactylopius coccus*). The results showed that indigo had the highest UV protection values.

Deodorizing Finishing

In terms of the younger generation's concern for health and cleanliness, advancements in textile performance, especially odor, antimicrobial qualities, or protection from ultraviolet radiation, have increased recently. Clothing odour is mostly caused by either human feces or bacterial colonies that have grown. Many significant studies have been published on the deodorizing qualities of fabrics made with natural colourants, to meet the growing demand for sanitary apparel among mature consumers. A comparison between the material dye and extracts of natural colourant revealed that pomegranate rind, coffee sludge, gardenia, and *Cassia tora L.* had the best deodorizing properties. When compared to undyed materials, fabrics dyed with gallnuts show superior deodorizing properties against ammonia, trimethylamine, and acetaldehyde.

Additionally, the dyed textiles demonstrated outstanding antibacterial action against *Klebsiella pneumoniae* and Staphylococcus aureus. Young-Hee Lee and colleagues colored cotton, silk, and wool fabrics with pomegranate (*Punica granatum*) extract for deodorizing purposes and achieved good results, with a range of 99 percent. Cotton textiles were pre- and post-mordanted with Cu (II) sulphate to deodorize ethyl mercaptan after being dyed with C. I. Direct Blue 200, a copper complex direct dye. Based on the findings, it was discovered that the impact of deodorization on copper ion uptake increased quadratically with every deodorization impact plotted against it. Therefore, by adhering to the correct dyeing process (optimized), both natural and synthetic dyes can be used for deodorizing and functionalizing fabrics [23].

Moth-Resistant and Insect-Repellent Textiles

The warmth, softness, and flame resistance of wool and other hair fibres make them ideal for use in carpets, shawls, and blankets. Textiles made of wool are especially vulnerable to moth and other insect infestations because of their high protein content. As insects, the protein in wool is consumed by moths as their larvae. Two common insects that prey on wool textiles are the clothes moth (*Tineola bisselliella*) and the carpet beetle (*Anthrenus verbasci*). Cyhalothrin, permethrin, hexahydro pyrimidine derivative, and DDT (Dichlorodiphenyltrichloroethane) are used as anti-moth finishing agents. Anthrenus verbasci, the larval form of the carpet beetle that feeds on protein fibres, was likewise prevented from feeding on wool fabric by the application of nano TiO_2 particles. Natural colorants might be the best substitutes for anti-moth finishing chemicals because they are all associated with environmental harm. For anti-moth finishing, Shakyawar *et al.* screened a variety of natural dye sources, including saffron flower waste, onion skin, henna, myrobolan, silver oak leaves, walnut, dholkanali, and yellow roots. They discovered that the best results came from pomegranate rind, walnut husk, and silver oak leaves [24]. Natural flavonoids, such as logwood, chestnut, fustic, and indigo, as well as natural quinones, such as cochineal, madder, and walnut, were used to dye wool to evaluate their antifungal properties against black carpet beetles. All the dyes, with the exception of indigo, increased the wool fabric's insect resistance; flavonoid dyes, however, had a less significant effect. Metallic mordants did not significantly affect insect resistance when all-natural dyes were used. It was discovered that anthraquinone dyes, such as those found in walnut, madder, and cochineal, were highly effective in repelling black carpet beetles from wool fabric.

CONCLUSION

Natural dyes are deeply ingrained in human culture and history. They were used by ancient people all over the world to commemorate customs, express their identity, and adorn their daily lives. Although synthetic dyes took their place in the 19th and 20th centuries, natural dyes have experienced a significant resurgence in recent years as consumers' demand for sustainable and environmentally friendly products has increased. Natural dyes may now be extracted, purified, and applied more easily thanks to modern science. We can obtain vibrant, long-lasting colors while using less water and avoiding harmful chemicals thanks to new technology. However, scientists have found that natural dyes are capable of much more than merely coloring textiles. They can combat infections, protect against damaging UV rays, and even aid in the treatment of illnesses like cancer. The importance of natural dyes is growing as the world seeks safer and more environmentally friendly alternatives in various sectors, including fashion, healthcare, food, and energy. They blend innovation and tradition, demonstrating that something ancient may still play a significant part in addressing today's problems. In conclusion, natural dyes are a wise and sustainable choice for the future, rather than merely a relic of the past.

ACKNOWLEDGMENTS

The authors wish to recognize the application of AI tools, like Grammarly, during the writing process to enhance the clarity and language of the manuscript. They have thoroughly reviewed and modified the content as necessary and accept full responsibility for the content of the chapters. Additionally, they would like to emphasize that AI was not utilized in the chapters to create or modify images, aid in data analysis, or contribute to their research.

REFERENCES

[1] Das S, Roy Maulik S. Recent Approaches and Advancements in Natural Dyes. Natural Dyes and Sustainability 2024; pp. 63-78.

[2] Alegbe EO, Uthman TO. A review of history, properties, classification, applications and challenges of natural and synthetic dyes. Heliyon 2024; 10(13): e33646. [http://dx.doi.org/10.1016/j.heliyon.2024.e33646] [PMID: 39040261]

[3] Yusuf M, Shabbir M, Mohammad F. Natural colorants: Historical, processing and sustainable prospects. Nat Prod Bioprospect 2017; 7(1): 123-45. [http://dx.doi.org/10.1007/s13659-017-0119-9] [PMID: 28093670]

[4] Harris S, Brøns C, Zuchowska M. Textiles in Ancient Mediterranean Iconography. Oxbow Books 2021.

[5] Melo, M.J., History of natural dyes in the ancient Mediterranean Civilization. Handbook of natural colorants, 2023: p. 3-26. [http://dx.doi.org/10.1002/9781119811749.ch1]

[6] Ahmed, H.E., History of Natural Dyes in North Africa_Egypt. Handbook of natural colorants, 2023: p. 33-41.

[7] Khalilu R, Areo MO. Technology and Adaptation in Yoruba Indigo Dyeing. International Journal of Textile and Fashion Technology 2015; 5(5): 5-18.

[8] Omowumi BJ, Benard SA, Afolabi OO, Fowotade AA, Olutunde OA. Camwood (Baphia nitida) alcoholic extract: A suitable counter stain for haematoxylin in the demonstration of liver and kidney histomorphology. Afr J Biotechnol 2018; 17(34): 1062-6. [http://dx.doi.org/10.5897/AJB2017.16287]

[9] Mikropoulou E, Tsatsaroni E, Varella EA. Revival of traditional European dyeing techniques yellow and red colorants. J Cult Herit 2009; 10(4): 447-57. [http://dx.doi.org/10.1016/j.culher.2009.02.003]

[10] Thadhani VM, Amaratunga G, Musharraf SG, Choudhary MI. Method of extraction of an effective tea dye powder from tea waste and application thereof on fabric and garments. WIPO (PCT) 2021; WO2021033173A2.

[11] Thadhani VM, Amaratunga G, Musharraf SG, Choudhary MI. Method for preparing effective dye using tea waste. NIPO (Sri Lanka) 2019. Patent No. 20656

[12] Thadhani VM., Amaratunga G, Musharraf SG, Choudhary MI. Method for Preparing Effective Dye using Tea Waste, National Intellectual Property Office, Sri Lanka, Patent No. 20656, (2019).

[13] Jordeva, S., *et al.*, Dyeing of textiles with natural dyes. Tekstilna industrija, 2020. 68(4): p. 12-21.

[14] Ferreira ESB, Hulme AN, McNab H, Quye A. The natural constituents of historical textile dyes. Chem Soc Rev 2004; 33(6): 329-36. [http://dx.doi.org/10.1039/b305697j] [PMID: 15280965]

[15] Prabhu K, Bhute AS. Plant based natural dyes and mordants: A Review. J Nat Prod Plant Resour 2012; 2(6): 649-64.

[16] Repon MR, Dev B, Rahman MA, *et al.* Textile dyeing using natural mordants and dyes: a review. Environ Chem Lett 2024; 22(3): 1473-520. [http://dx.doi.org/10.1007/s10311-024-01716-4]

[17] Che J, Yang X. A recent (2009–2021) perspective on sustainable color and textile coloration using natural plant resources. Heliyon 2022; 8(10): e10979. [http://dx.doi.org/10.1016/j.heliyon.2022.e10979] [PMID: 36281409]

[18] Richhariya G, Kumar A, Tekasakul P, Gupta B. Natural dyes for dye sensitized solar cell: A review. Renew Sustain Energy Rev 2017; 69: 705-18. [http://dx.doi.org/10.1016/j.rser.2016.11.198]

[19] Latos-Brozio M, Masek A. The application of natural food colorants as indicator substances in intelligent biodegradable packaging materials. Food Chem Toxicol 2020; 135: 110975. [http://dx.doi.org/10.1016/j.fct.2019.110975] [PMID: 31747619]

[20] Vashishth, P., *et al.*, Natural Dyes as Corrosion Inhibitors, in Natural Dyes and Sustainability. 2024, Springer. p. 265-286.

[21] Shahin A, Shreadah M, El-Nagar K, Nabil-Adam A, Osman H. Application of printing and finishing cotton fabrics with natural dyes (a review). Egypt J Chem 2022; 65(11). [http://dx.doi.org/10.21608/ejchem.2022.121633.5451]

[22] Grifoni D, Bacci L, Zipoli G, Albanese L, Sabatini F. The role of natural dyes in the UV protection of fabrics made of vegetable fibres. Dyes Pigments 2011; 91(3): 279-85. [http://dx.doi.org/10.1016/j.dyepig.2011.04.006]

[23] Mishra A, Gautam S. Application of natural dyes for herbal textiles. Chemistry and technology of natural and synthetic dyes and pigments. IntechOpen 2020. [http://dx.doi.org/10.5772/intechopen.93183]

[24] Shakyawar DB, Raja AS, Kumar A, Pareek PK. Antimoth finishing treatment for woollens using tannin containing natural dyes. Indian J Fibre Text Res 2015; 40(2): 200-2.

CHAPTER 2

Sources of Natural Dyes

Naheed Akhtar[1,*], **Syed G. Musharraf**[1,2,3] and **Vinitha M. Thadhani**[4]

[1] *H.E.J. Research Institute of Chemistry, International Center for Chemical and Biological Sciences, University of Karachi, Karachi, Pakistan*

[2] *Dr. Panjwani Center for Molecular Medicine and Drug Research, International Center for Chemical and Biological Sciences, University of Karachi, Karachi, Pakistan*

[3] *School of Chemistry and Pharmaceutical Sciences, Guangxi Normal University, Guilin, Guangxi, China*

[4] *Ceylon Ayurveda Pvt. Ltd, Havelock Road, Colombo, Sri Lanka*

Abstract: This chapter describes the origins of natural dyes, their diverse varieties, and the mechanisms by which they act. Minerals, plants, animals, and even microscopic living things like fungi and bacteria can be used to make natural dyes. Every source has its benefits and drawbacks, offering a range of perspectives.

This chapter categorizes various natural dyes—including flavonoids, tannins, carotenoids, and anthraquinones—by their chemical composition, explaining how these structures determine the resulting color's hue and intensity. Another approach to categorizing dyes is by application method; some require specific fixing agents or mordants, while others can be applied straight.

A straightforward explanation of the science underlying dyes' ability to adhere to textiles and how their molecules impart color is provided. This chapter provides a comprehensive overview of natural dyes, demonstrating how their diverse sources, structures, and applications make them valuable for textiles and a wide range of other applications.

Keywords: Disperse Dye, Henna, Indigoid, Lichens, Tyrian purple, Tannins.

INTRODUCTION

For millennia, people have used natural dyes to color textiles, food, and other objects. They come from naturally existing sources, such as microorganisms, minerals, plants, and animals. These dyes are more environmentally friendly than synthetic ones, and their popularity is rising as consumers want more sustainable

* **Corresponding author Naheed Akhtar:** H.E.J. Research Institute of Chemistry, International Center for Chemical and Biological Sciences, University of Karachi, Karachi, Pakistan; E-mail: naheedakhtar41@yahoo.com

and eco-friendly products. The most prevalent source of natural dyes is plants. Their colors are obtained from leaves, roots, flowers, fruits, seeds, and bark. For instance, turmeric imparts a yellow hue, and indigo is derived from indigo plants. Depending on the growth environment, their color may fade more quickly, and the dye production may change. They are generally accessible, biodegradable, and renewable.

Natural dyes are also produced by certain insects and marine life. Carmine is a vivid red dye produced by cochineal insects. Sea snails are the source of Tyrian purple. Although these colors are rich and durable, some users may have ethical questions about them. These are derived from soil and rocks with natural colors. Examples are blue from azurite, yellow from ochre, and red from cinnabar. They are less colorful than plant or animal dyes, but they are more stable and long-lasting. Some fungi and bacteria have the ability to create colors spontaneously. These can be cultivated in laboratories and utilized for textiles and other purposes in the future; however, research is currently ongoing. Flavonoids, tannins, and carotenoids are examples of natural colors that can be categorized based on their chemical structures. These structures determine each dye's color and fabric-retention properties. Lastly, dyes can be categorized based on their intended function. Some can be used straight away, while others require a mordant to fix the color. Selecting the appropriate dye for each material is made easier by being aware of these variations.

SOURCES OF NATURAL DYES

Dyeing is as ancient as human culture. Natural dyes were exclusively employed to color fabrics from antiquity until the eighteenth century. Natural resources are the source of natural dyes, as the name implies. To color a variety of textile fabrics, coloring agents derived from natural resources of plant, animal, mineral, and microbiological origin were employed. Traditional methods of natural dyeing were practiced in several parts of the world, utilizing local natural resources [1, 2].

Types of Natural Dyes

Natural dyes have been categorized according to their chemical structure, application techniques, and sources of production or origin [3]. The classification chart for natural colorants is shown in Fig. (**1**).

Based on Origin Sources

Even though plants are the primary source of natural dyes, these substances can be roughly categorized as plant, animal, mineral, or microbial colors depending on where they come from. Many researchers documented and compiled traditional dyeing techniques in many countries, as well as gathered knowledge from ancient texts, as interest in natural dyes rose. Below is a brief summary of the possible dye resources arranged by their place of origin [4]. Fig. (**2**) shows the classification of natural dyes based on their origin sources.

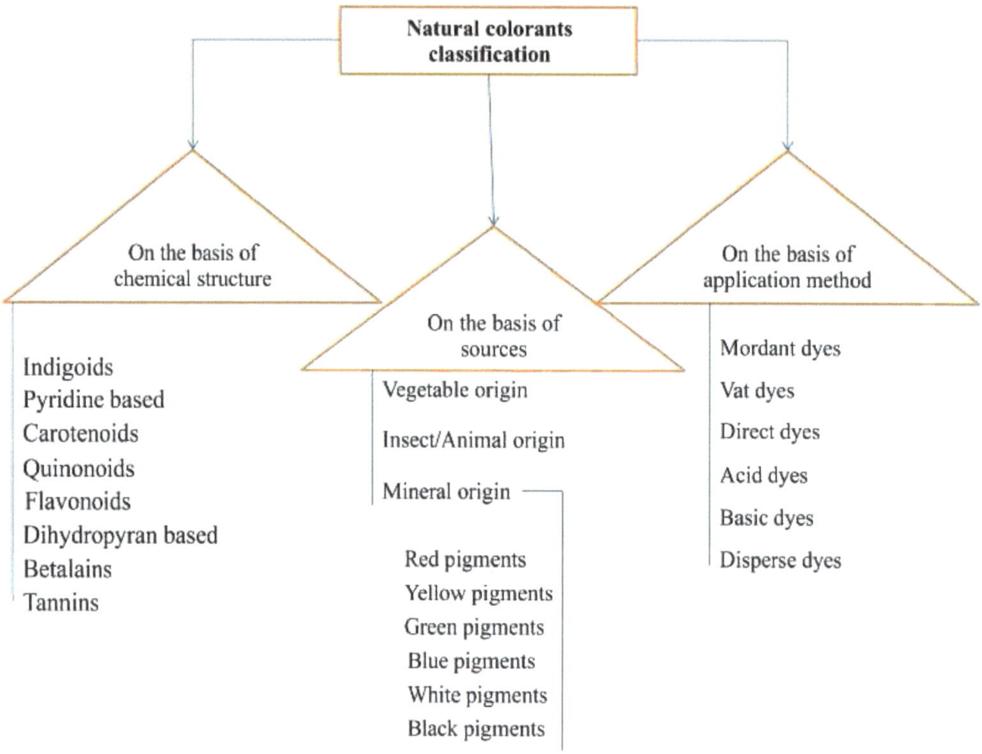

Fig. (1). Classification chart for natural colorants.

Plants

The vast majority of natural dyes have historically been extracted from plants. Roots, leaves, twigs, stems, heartwood, bark, wood shavings, flowers, fruits, rinds, hulls, husks, and other plant parts can all be used as natural dyes [3].

In many civilizations, plant-based dyes have long been used to color textiles, crafts, and artwork. These dyes come from a variety of plant parts, such as the roots, leaves, stems, flowers, and fruits, as well as from the biomass of the plants that yield a broad spectrum of hues. Numerous plant species are well known for their vivid and unique hues, which have been used to dye clothing, cosmetics, and other items. These plants' active chemical components are frequently what give them their hues. Indigo, madder, turmeric, henna, logwood, onion, black walnut, marigold, woad, banana, pot marigold, roselle, kola nut, annatto, yam, rosary pea, mauve, and tea waste are a few examples of these plants [5].

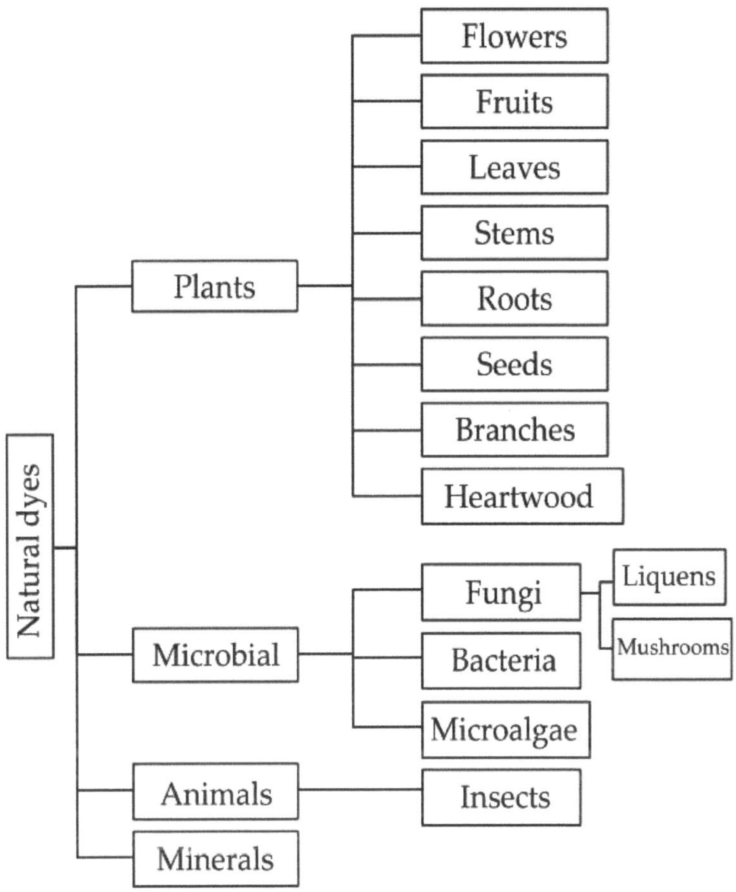

Fig. (2). Classification of natural dyes based on origin sources.

For thousands of years, indigo has been utilized as one of the earliest known natural plant-based dyes. Fermentation techniques are used to extract the indigo plant's blue hue. Many societies around the world have used this method. Likewise, red and orange dyes can be obtained from the root of madder (*Rubia tinctorum*). Ancient civilizations like Egypt and Greece employed it, and it has been cultivated for ages. Bright yellow and orange hues are produced by the spice turmeric (*Curcuma longa*), which has potent dyeing qualities. Asian cultures have utilized it in traditional medicine and for textile dyeing. Curcuminoids are the primary source of turmeric's vibrant yellow color, with curcumin serving as the primary active ingredient that gives it its distinctive hue. Similarly, henna

(*Lawsonia inermis*) is well-known for its application in hair and body art. Lawsone, a naturally occurring pigment in henna plant leaves, is what gives henna its reddish-brown hue. *Haematoxylum campechianum*, or logwood, yields a variety of hues, including rich blues and purples. It is derived from the logwood tree's heartwood and has long been utilized to make premium colors.

Beautiful hues of brown and yellow can be made from onion peels. This exemplifies a method of natural dyeing with kitchen waste. In North America, black walnut (*Juglans nigra*) hulls have long been utilized for their rich dark color. Europe uses a bright yellow dye made from the yellow blooming plant Weld (*Reseda luteola*). Marigold (*Tagetes spp.*) petals are used to make vibrant orange and yellow colors. Traditional medicine and dyeing are two common uses for them. Similar to indigo, woad plants (*Isatis tinctoria*) were used in Europe to make blue dyes. Chlorophyll, a pigment necessary for photosynthesis, gives the banana's (*Musa sapientum*) leaves and fruit their deep green hue. The presence of carotenoid pigments, such as lutein and zeaxanthin, is responsible for the pot marigold's (*Calendula officinalis*) vibrant orange and yellow petals. The presence of anthocyanin pigments, specifically cyanidin-3-sambubioside, gives the calyces of roselle (*Hibiscus Sabdariffa*) their deep red color, while tannins, specifically catechins and proanthocyanidins, give the kola nut (*Cola acuminata*) its brown hue and contribute to its astringent qualities. While yam flesh (*Dioscorea rotendata*) ranges in color from purple to yellow due to the presence of anthocyanins and carotenoids, respectively, anatto seeds (*Bixa orellana*) produce a yellow to orange color due to the presence of carotenoid pigments, including bixin and norbixin. Abrin, which gives the rosary pea (*Abrus precatorius*) seeds their vibrant red color, also gives them strong cytotoxic effects. It is difficult to find in the literature the precise component that gives mauve (*Annickia chlorantha*) its particular tint. However, as the majority of other plants frequently get their colors from these pigments, the mauve gets its color from flavonoids, anthocyanins, and carotenoids. Some plants that are investigated for their ability to produce dyes and colors are shown in Table **1**.

Table 1. Plants explored for dyes and colors produced.

Plants	Parts Used	Colors
Abies spectabilis (Himalayan Fir)	Cone	Purple/Violet
Abrus precatorius (Rosary Pea)	Flower	Red
Acacia nilotica (Gum arabic, Babool)	Leaves/Bark	Yellow/Brown
Adhatoda vasica Nees (Malabar nut)	Leaves	Yellow
Aegle marmelos (Bael)	Fruit rind	Yellow
Ageratina adenophora (Sticky snakeroot)	Leaves	Yellow

(Table 1) cont.....

Plants	Parts Used	Colors
Ageratum conyzoides (Goat weed)	Whole plant	Yellow
Allium cepa (Onion)	Fruits/Leaves	Yellow
Aloe barbadensis L. (Aloe vera)	Whole plant	Red
Annickia chlorantha (Mauve)	Stem bark	Brown
Azadirachata indica (Neem)	Bark	Brown
Barleria prionitis L. (Porcupine flower)	Flower	Yellow
Berberis aristata (Indian barberry)	Bark	Yellow
Betula utilis (Bhojpatra)	Tree gum	Brown
Bixa orellana (Annatto)	Fruit	Reddish Orange
Butea monosperma (Palash)	Flower	Yellow/Orange
Calendula officinalis (Pot Marigold)	Flower	Yellow
Camellia sinensis (Tea)	Leaves	Brown
Capsicum annum L. (Chili)	Fruits	Red
Cassia fistula L. (Golden shower)	Bark	Brown
Castanopsis indica (Indian chestnut)	Bark	Brown
Cedrela toona (Cedrela)	Flower/Leaves	Yellow/Red
Chromolaena odorata (Siam weeds)	Whole plant	Yellow
Cola acuminata (Kola nut)	Nut	Brown
Crocus sativus L. (Saffron)	Flower	Yellow/Orange
Curcuma longa (Tumeric)	Root/Rhizome	Orange/Yellow
Dioscorea rotundata (Yam)	Tuber	Brown
Embelia ribes (Embelia)	Fruits	Red
Eugenia jambolana L. (Jambolan)	Bark/Leaves	Red
Gardenia jasminoides (Cape jasmine)	Fruits	Yellow
Gossypium herbaceum (Levant cotton)	Flowers	Yellow
Hibiscus sabdariffa (Roselle)	Calyx	Red
Lawsonia inermis (Henna)	Flower	Reddish brown
Musa sapientum (Banana)	Stem bark	Yellow/orange
Senegalia catechu (Catechu)	Bark	Brown/Black

Advantages of Plant-based Dyes

Plant-based dyes offer many benefits that lead to their use in the food, cosmetics, textile, and pharmaceutical industries. First of all, because they can be produced and harvested without the use of artificial chemicals, they are sustainable and renewable. They are a sustainable option because they are typically derived from plant fruits, roots, leaves, and flowers. Second, they often decompose spontaneously without adding to environmental damage because they are biodegradable. Additionally, because plant-derived colors include bioactive components, they have health-promoting qualities. For example, several plant pigments contain antibacterial and antioxidant qualities that are beneficial in a range of medical applications [6].

Challenges of Plant-based Dyes

Plant dyes have certain drawbacks despite their benefits. The variety of colors that can be obtained from the same plant source is one of the main obstacles. Because of this, it is challenging to identify and measure the distinct pigments in a given plant as well as to achieve uniform color hues. The intricacy of plant extracts, which can include a variety of chemicals in addition to color pigments, presents another difficulty in their separation and analysis. Furthermore, the extraction and preparation of plant dyes can be laborious and call for specific skills and tools. Plant dyes have the tendency to deteriorate over time, which can shorten the lifespan of dyed goods. Plant-based dyes' commercial viability and broad use are typically constrained by these issues [5].

Animal

Colorants made from animal parts such as scales, bones, shells, and secretions are known as animal-based dyes. Many cultures have been dyeing fabrics, ceramics, and other things using these dyes for thousands of years. Historical records show their significance in trade and cultural customs, and they offer a variety of colors and shades, from earthy tones to vivid hues. Cochineal, Tyrian purple, sepia, oyster purple, guanine, bone black, and squid ink are a few examples of animal-based colors. However, overexploitation of some of these sources, such as Murex snails, has raised ecological issues.

Cochineal (Carmine)

The crimson dye cochineal is made from the dried bodies of female Dactylopius coccus Costa insects, which eat cacti, particularly Opuntia. After the insects are crushed, dried, and harvested, the red pigment is removed. The primary color that may be extracted from cochineal insects is carminic acid. Throughout Central and South America, cochineal has been used for generations and was a significant export during colonial times. People still use this natural food and cloth dye nowadays. Cosmetics, food, and pharmaceutical applications are its primary uses; textile and plastic applications are also possible [7].

Tyrian Purple

Some sea snails, including the *Murex brandaris* and *Murex trunculus*, secrete a substance called Tyrian Purple, sometimes referred to as royal purple. After crushing the snails, a chemical process is triggered, producing a purple color. Ancient societies held Tyrian Purple in great regard, especially in Greece and Rome. It is a costly dye connected to the aristocracy.

Sepia and Shellfish Purple

The ink sac of the cuttlefish, Sepia officinalis, is the source of sepia dye. Brown pigment is produced by processing and removing the ink sac. Historically, sepia has been used as a dye and an ink. Shellfish *Hexaplex trunculus* is one of several mollusk species from which the purple color is extracted. Made from the mollusk's secretions, it was prized as a luxury item by the ancient Greeks and Phoenicians.

Guanine, Bone Black, and Squid Ink

The scales of fish, especially herring and sturgeon, are used to make guanine. The silvery pigment is extracted by crushing, processing, and gathering the scales. Guanine is used to produce a shimmering look in nail polish, cosmetics, and other applications. Charred animal bones, typically those of cattle, are used to make Bone Black. High temperatures are applied to the bones, producing a black carbon pigment. In the past, bone black was utilized as a black pigment for dyeing, ink, and painting. The ink sac of squid, including Loligo and Sepia species, is used to make squid ink. Harvested, the ink sac is utilized as a natural black pigment in cooking to provide food color and taste.

Mineral-Based Dyes

Colorants made from naturally occurring minerals are referred to as mineral-based dyes, inorganic dyes, or mineral pigments. The excellent chemical stability and resistance to fading or chemical alteration under various environmental circumstances are characteristics of these pigments, which are made up of inorganic compounds. For millennia, they have been used in a variety of fields, such as ceramics, art, textiles, and cosmetics. Ultramarine blue, which comes from the stone lapis lazuli, is among the most well-known instances of a mineral-based dye. Because of its intense blue hue and fading resilience, this pigment has been utilized since ancient times. Ochre, a naturally occurring pigment made from iron oxide and available in a variety of hues such as yellow, red, and brown, is another example. The longevity and permanence of mineral-based dyes are their main advantages. Mineral pigments are highly resistant to elements like light, heat, and moisture, in contrast to organic dyes, which are frequently made from plant or animal sources and may fade or degrade with time. Because of this feature, they are especially well-suited for uses where colorfastness and durability are essential. Mineral-based dyes have been crucial to understanding ancient civilizations and their artistic practices in the fields of historical studies and art conservation. For example, studying the colors used in cave paintings reveals information on the resources early humans had at their disposal and their cultural customs. Because of their durability and non-toxicity, mineral-based dyes are now used in a variety of industries, including cosmetics, where they are preferred over synthetic organic dyes. For instance, common mineral-based pigments used in cosmetics that offer a variety of natural colors are titanium dioxide and iron oxides [5]. Natural colorants from mineral origin can further be classified by their colors.

Red Pigments

Examples of red colors that come from minerals include cinnabar, red ochre, red lead, and realgar. A typical source mineral for refining elemental mercury, cinnabar, sometimes called vermilion, is a brilliant scarlet to brick-red type of mercury sulphide (HgS) that is used directly as a coloring pigment. The natural earth color known as red ochre (Geru in Hindi) contains both hydrated and anhydrous iron oxide ($Fe_2O_3{\cdot}nH_2O$). Although red ochre is not as vivid as cinnabar, it comes in a variety of shades, from yellow to deep orange to brown. The compound red ochre is highly stable and unaffected by light, acids, or alkalies. Its crude variant is washed to produce fine red ochre. Monks paint their robes with red ochre. The vivid red or orange crystalline or amorphous pigment known as red lead (Sindur in Hindi) (Pb_3O_4 or $2[PbO]{\cdot}[PbO_2]$) is widely used in Indian art. Often referred to as ruby sulphur or ruby of arsenic, realgar (α-As_4S_4) is an arsenic sulphide mineral that is found in conjunction with orpiment (As_2S_3), another arsenic mineral. Both are arsenic sulphides, although they are dangerous and haven't been used in paintings very often.

Yellow Pigments

Because of their range of yellow hues, Yellow Ochre (Ram Raj), Raw Sienna, Orpiment, and Litharge (Massicot) are categorized as yellow pigments. The presence of several hydrated forms of iron oxide, especially the mineral limonite ($Fe_2O_3{\cdot}H_2O$), is what gives the yellow ochre its color. Selection, grinding, washing, and ligation are the steps used to prepare the pigment from natural soil. Iron oxide and manganese oxide are found in the Sienna (Siena earth) class of earth pigments, which includes raw sienna. It was the first pigment used in human cave drawings, along with ochre and umber. Because of its considerable transparency, it is utilized as a glaze in paintings. Orpiment (called "Hartal" in Hindi) is a mineral of arsenic sulphide that has a rich, deep orange-yellow hue. In terms of chemistry, it is arsenic's yellow sulphide (As_2S_3). It has been used to tint paper yellow in addition to being a pigment. Additionally, this procedure gives paper an insecticidal quality. White lead is gently roasted to create litharge (Massicot), a naturally occurring secondary mineral form of lead oxide (galena). After decarboxylation and dehydration, white lead, which is chemically lead carbonate ($2PbCO_3 \cdot Pb(OH)_2$), turns into a pale yellow powder that is lead monoxide when heated to roughly 300 °C (PbO).

Green Pigments

Vedgiris, Malachite, and Terre-Verte (Green Earth). Since ancient times, terre-verte has been the most commonly utilized of them. Although other minerals are probably present, green earth is a mixture of hydrosilicates of Fe, Mg, Al, and K (gluconite and celadenite). Depending on where it comes from, green soil has different colors. Light and chemicals do not affect the colors, which range from yellow-green to greenish grey. Malachite, which has the chemical formula $Cu_2(OH)_2CO_3$, is a copper carbonate hydroxide mineral. The monoclinic crystal system is how this opaque, green-banded mineral crystallises. Vedgiris was a popular pigment used in miniature paintings and paintings during the Mughal era. It is made by applying vinegar to copper foils and is the typical acetate of copper [$Cu(CH_3COO)_2$]. The resulting pigment is deep green and extremely brilliant. Its drawback is that, if not used cautiously, it can burn paper or textiles.

Blue Pigments

Blue pigments include azurite and ultramarine blue. The semi-precious stone lapis lazuli is the source of ultramarine blue, also known as Lajward in Hindi, a deep blue pigment. In India, it has been incorporated into tiny paintings. In the 14th and 15th centuries, lapis lazuli was brought to India from Afghanistan. The weathering of copper ore deposits produces the soft, deep blue pigment known as azurite [$Cu_3(CO_3)_2(OH)_2$]. Indian paintings hardly ever utilized this pigment, although Chinese paintings did. It has been claimed, nevertheless, that this mineral is present in Indian copper ores.

White Pigments

White pigments include zinc white, white lead, and chalk (white lime). One of the forms of calcium carbonate is chalk ($CaCO_3$). Paintings have made considerable use of it. Chalk has been used as a pigment since ancient times and is found in limestone deposits. Conch shell white is thought to have unique qualities and was favoured by painters in India. White lead ($PbCO_3$) is a complicated salt that has hydroxide and carbonate in it. In the past, it was a component of lead paint. The mineral Cerussite can be found in nature. However, White lead is typically made artificially. Another crucial pigment used in painting is zinc white (ZnO), also known as Safeda in Hindi. Before its introduction in Europe, zinc white was used as a pigment in India, according to archaeological data. Titanium white, barium white, and talc are further white pigments. Titanium White, also known as titanium dioxide (TiO_2), is used as a delustrant in textiles.

Black Pigments

The list of black pigments includes Charcoal Black, Lamp Black, Ivory Black, Bone Black, Graphite, Black Chalk, and Terre-noire (Black Earth). Traditionally, black pigment has been made from well-ground charcoal. In India, black pigment is made by powdering charcoal made from tamarind tree twigs and wood that has been burned in a covered pot. Almond and coconut shells are among the various materials that, once charred, were utilized to make black pigment. The resulting charcoal is soft and produces a fine, uniform black pigment. "Kajal," which is made by burning oil in a lamp and spreading the soot onto an earthen bowl, is by far the most significant black product used in India. Charring ivory slices in a closed clay pot, followed by grinding, washing, and drying the black residue, is how ivory black is made. So prepared, the black is intense. Due to concerns about the environment and animal welfare, it is currently not preferred. Animal bones are burned in confined earthen pots to create bone black. Though used as a replacement, it lacks the intensity of ivory black. Writing materials have been made from powdered graphite, a mineral that may be found across India. The pigment is drab and grey. But rather than being used for painting, it has primarily been employed for drawing. Black clay used for pottery and paintings is called black chalk. Black clay is the same as terre-noire. It is a combination of clay with calcium, iron, and manganese carbonate [8].

Primarily used for painting, handicrafts, antiques, and the restoration of cultural artefacts, mineral dyes are derived from natural minerals. Cinnabar (HgS) has been used as a red dye from ancient times. It was frequently used in medieval manuscripts with colored drawings or patterns, as well as in the art of ancient Rome, and for decoration. Local ores in the southern Pyrenees are the source of the light-blue mineral aerinite. Romanesque wall paintings in Andorra and Catalonia, Spain, employed blue dye, which can be made from aerinite. In the past, lapis lazuli-based natural ultramarine (UMB) dyes have been employed. As one class of mineral dyes, the UMB dyes are distinguished by their sodalite structure and the colorful sulphur species that are contained within the chromophores. UMB dyes have the general formula $Na_8[AlSiO_4]_6[S_2S_3]_2$. Blue, green, red, and black hues were extracted from rocks or minerals by ancient painters, potters, and craftspeople. Consequently, food additives are less appropriate for usage in art and architecture than mineral dyes [9].

Microbial/ Microorganism

The most widely dispersed living things are communities of microorganisms. They are an essential component of the biosphere and ecosystem and have close relationships with plants, animals, and other microbes through parasitism, symbiosis, and saprophytic states. A significant source of natural colors is microorganisms, such as bacteria, fungi, and some algae. Carotenoids (β-carotene, canthaxanthin, and astaxanthin), bacteriochlorophylls, flavonoids, indigoids, melanin, pheomelanin, prodigiosin, violacein, glaukothalin, phycocyanin, and xanthomonadin are among the extraordinary variety of microbial pigments that are currently available in a variety of environments. Potential dye-producing fungi have been identified, and some of these include the following: Epicoccum nigrum, Lecanicillium aphan-ocladii, Fusarium sp., Aspergillus niger, *Aspergillus versicolor*, *Monascus sp.*, *Trichoderma viride*, *Penicillium purpurogenum*, *Aspergillus sydowii*, *Aspergillus aureolatus*, *Aspergillus keveii*, *Penicillium fungus*, *Penicillium chermesinum,* Fusarium sp. Some fungal dyes, like Monascus pigments, Arpink red from *Penicillium oxalicum*, riboflavin from *Ashbya gossypii*, and β-carotene from *Blakeslea trispora*, have already been utilized as food colorants in the market. Worldwide, monascus pigment has been used extensively as food coloring since the 1880s. Monascus pigment has been used extensively in the food industry throughout Asia, particularly in Chennai and Japan. Fungi provide significant economic and productivity benefits over plants and mammals. Fungal dyes, for instance, can grow quickly and easily in a low-cost culture medium and are not affected by the season. Fungal dyes in particular exhibit good stability, solubility, a variety of color hues, and ease of processing. Since the majority of microbial dyes are safe for the environment and nontoxic to humans, they can be employed as food additives and in the medical field. Microbial dyes are predicted to replace synthetic colorings. An excellent and accessible source of natural dyes might be fungi [9].

Based on Chemical Structure

The most acceptable and commonly used method of classifying natural dyes is based on their chemical structure, which makes it easy to identify colors that belong to a specific chemical group with specified characteristics [3].

Indigoids

The oldest natural dyes utilized by human societies are indigoids, which include indigo and tyrian purple. They are arguably the most significant group of colors.

One of the most significant and precious coloring materials is natural indigo, a dye with a unique blue hue that has a long history. *Perisicaria tinctoria*, *Isatis tinctoria*, *Polygonum tinctorium* (dyer's knotweed), and Indigofera spp. (*Indigofera tinctoria*) are the sources of indigo (woad). However, a significant portion of indigo today—thousands of tons annually—is synthetic. Tyrian purple (C.I. 75800), also called Tyrian red, royal purple, or imperial purple, is a reddish-purple natural dye that contains bromine and is made from the hypobranchial glands of a number of marine predatory sea snails in the Muricidae family. This dye offers outstanding light-fastness qualities [3, 10].

Indigo has been used as a dye for around four hundred years. Among a number of categories, it is the most important natural dye. Two essential natural colors, indigo and Tyrian purple, both contain an indigoid structure. Indigo blue was one of the first and best-known colorants. Indigo-dyed blue jeans continue to be a popular color, as seen by their widespread use. Before its potential as a blue dye was discovered, indigo was used as a black dye. The molecular structure of indigoid dyes, an aromatic component, is formed by the combination of two distinct indoxyl compounds with nearly identical chemical components found in *Indigofera suffruticosa* and *Isatis tinctoria* var. Since indigo is a water-insoluble dye, it must first be converted from its water-insoluble form to its water-soluble leuco form via a widely used procedure known as the reduction process before being applied to textile fabrics. *Strobilanthes flaccidifolius, Isatis tinctoria, Wrightia tinctoria, Indigofera tinctoria*, the genus *Purpura*, and the genus *Murex* are the sources of indigoid dyes [11]. Tyrian purple and Indigo blue are two significant colors in this class. It appears in the plant as a glucoside indicator. Woad is another blue dye that belongs to the same chemical class [12]. Fig. (**3**) shows the chemical structures of Tyrian purple (a) and Indigo blue (b).

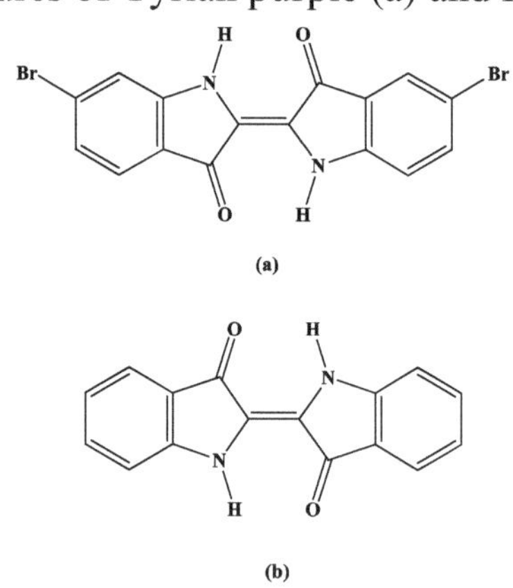

Fig. (3). Chemical structures of Tyrian purple (a) and Indigo blue (b).

Quinonoids

Quinonoids, which range in color from yellow to red, are abundant and widely distributed in nature. Compared to other plant pigments, the chemical structures of naturally occurring quinones are the most varied. These colors are further categorized as benzoquinones, α-naphthoquinones, and anthraquinones based on their chemical structure (Fig. **4**). *Carthamus tinctorius* (Safflower), *Choloraphora tinctoria* (Gaudich), *Lawsonia inermis/Lawsonia alba* (Henna/Mehendi), *Juglans regia* (Walnut), *Plumbago capencis* (Chitraka/Chita), *Drosera whittakeri* (Sundew), *Tabebuia avellanedae* (Taigu/Lapachol), *Alkanna tinctoria* (Ratanjot/Alkanet), *Lithospermum erythrorhizon* (Tokyo Violet/Shikone), *Dactylopius coccus* (Cochineal), *Kermes vermilio/Coccus ilicis, Laccifer lacca/Kerria lacca/Coccus lacca, Rubia tinctorum, Rubia cordifolia* (Indian Madder), *Rheum emodi* (Himalayan rhubarb), *Oldenlandia umbellata* (Chay Root), and *Morinda citrifolia* (Al/surangi/ach) are the natural resources for quinonoids class; subclass anthraquinonoids and naphthoquinonoids [3].

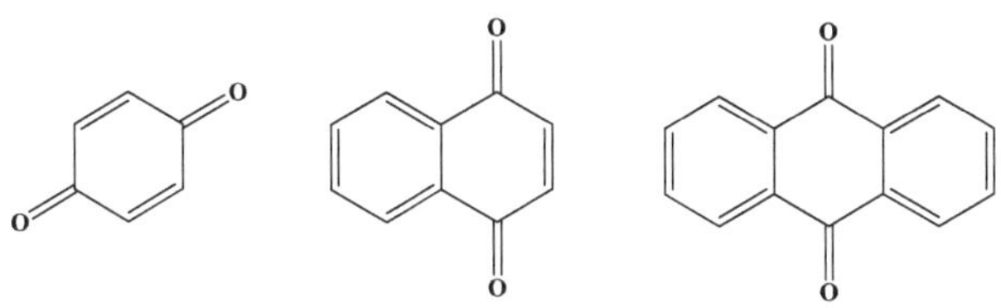

(a) (b) (c)

Fig. (4). Chemical structures of benzoquinone (a), 1,4-naphthoquinone (b), anthraquinone (c).

Anthraquinones

This category includes a wide variety of colorants, especially those that produce red color. In addition to having good color fastness to light, these also have good wash fastness when metal salt compounds are utilized. The most common color in this group is alizarin, a natural hue derived from European madder (*Rubia tinctorum*). Other sources of anthraquinone dyes include Indian madder (*Rubia cordifolia*), noni (*Morinda citrifolia*), cochineal (*Dactylopius coccus*), lac (*Laccifer lacca*), and so on [11]. This class of dyes is derived from plants and insects and has an anthraquinone molecule. This class is the only one with the red hue. Examples include cochineal, madder, lac, and kermes [12].

Alpha-naphthoquinones

Like anthraquinone dyes, these dyes produce red, orange, or reddish brown colors. The most well-known members of this group are juglone (5-hydroxy-1,4, naphthoquinone) and lawsone (2-hydroxy-1, 4-naphthoquinone). The lawsone class, which includes henna (*Lawsonia inermis*), is widely grown in Bangladesh,

India, Australia, and Africa. Henna-derived dye is used on cotton, wool, silk, and even polyester. The outer shell of a walnut is an example of the juglone class (*Juglans regia*) [11, 12].

Flavonoids

With hues ranging from mild yellow (isoflavones) to deep yellow (chalcones, flavones, flavonols, and aurones), orange (aurones), red, and blue, flavonoids offer the widest variety of plant dyes (anthocyanins). Flavonoid dyes can be obtained from a variety of plants, including Butea monosperma/Butea frondosa (Flame of the Forest/Palas), *Mallotus philippinensis* (Kamala), *Bignonia chica/Arrabidaea chenna* (Carajuru/Puca), *Commelina communis, Myrica esculenta* (Kaiphal), *Artocarpus heterophyllus/Artocarpus integrifolia* (Jackfruit), *Myrica esculenta (Kaiphal), Datisca cannabina* (Hemp), *Delphinium zalil* (Yellow Larksur), *Gossypium herbaceum*, and (Red Sandalwood) [3].

Flavonoids, also known as colorific groups, are phenolic chemicals that are produced by plants. As anti-inflammatory, anti-allergic, anti-tumor, antiviral, anti-diabetic, cardiovascular protective, antioxidant, and antibacterial agents, flavonoids are utilized in medicine. Flavones, flavonols, isoflavones, anthocyanins, aurones, and chalcones are among the various classes of flavonoids (Fig. **5**). The flavonoid group produces a variety of hues, including orange, red (anthocyanins), blue (anthocyanins), pale yellow (isoflavones), and deep yellow (chalcones, flavones, flavonols, and aurones). Weld (*Reseda luteola*), onion peel (*Allium cepa*), tickseed (*Coreopsis lanceolata*), red sandalwood (*Pterocarpus santalinus*), kaiphal (*Myrica esculenta*), hemp (*Datisca cannabina*), kamala (*Mallotus philippinensis*), carajuru (*Arrabidaea chenna*), and other plant sources are sources of flavonoid dyes [11, 12].

Tannins

The majority of the vegetable kingdom contains astringent substances called tannins. Fruit, pods, plant galls, leaves, bark, wood, and roots are just a few of the plant elements that contain tannins. Water-soluble phenolic compounds with molecular weights ranging from 500 to 3000 are known as tannins. Tannins are often divided into two categories: condensed tannins and hydrolyzable tannins, such as pyrogallol and proanthocyanidins (Fig. **6**). Gallotannis and ellagitannins are polyesters with a sugar moiety and organic acids that hydrolyze to produce galllic acid and ellagic acid, respectively. The main function of tannins is to preserve leather. Inks, glues, stains, and mordants all include tannins. In surface water treatment, tannins are also utilized to remove heavy metals. Because they increase fibers' affinity for various dyes, tannins are crucial when using natural dyes. It produces a variety of hues, including yellow, brown, grey, and black, when combined with various natural dyes. Plants that contain tannins include *Quercus infectoria* (Gallnut), *Terminalia chebula* (Harda), *Punica granatum* (Pomegranate/Anar), and *Acacia catechu* (Cutch) [3].

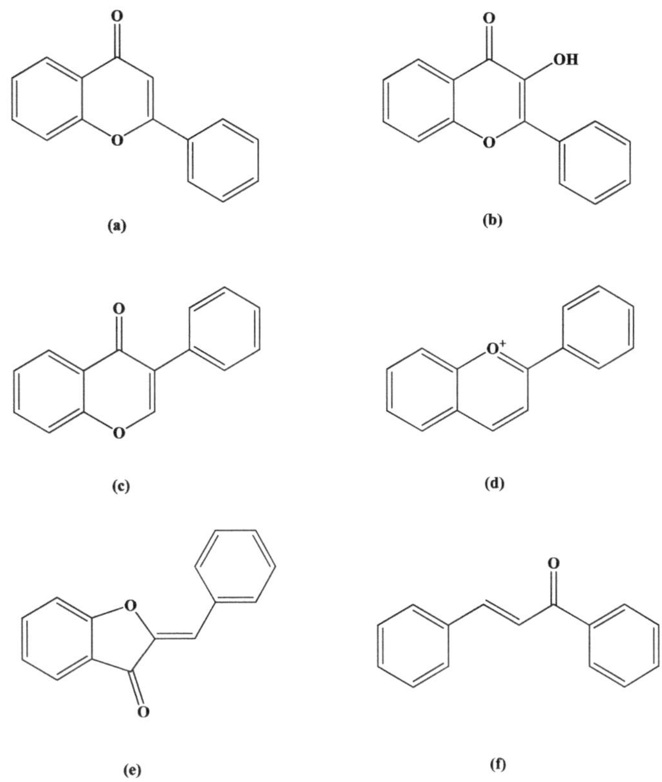

Fig. (5). Chemical structures of flavones (a), flavanols (b), isoflavones (c), anthocyanins (d), aurones (e), and chalcones (f).

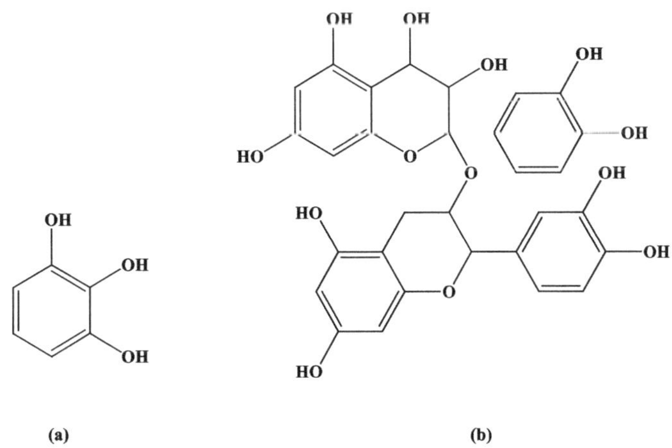

Fig. (6). Chemical structures of pyrogallol (a) and proanthocyanidins (b).

Tannins are polyphenolic compounds used for coloring that can be found in many natural sources. They can be separated into two categories: condensed tannins and hydrolyzable tannins. The molecular weights of condensed tannins are typically greater (1000–20,000 Da) than those of hydrolyzable tannins (500–3000 Da). The phenolic hydroxyl groups in them enable them to cross-link proteins and other big molecules efficiently. Plant materials, including leaves, bark, fruit, wood, roots, galls, and so forth, are used to extract vegetable tannins. Tannins produce black, grey, brown, and yellow hues. They also increase the fibers' receptivity to different colors. Tannins are mostly used for staining, mordanting, glue manufacturing, and leather retention. There are two types of tannins: condensed tannins (proanthocyanidins) and hydrolyzable tannins (pyrogallol, gallotannins, and ellagitannins). Harda (*Terminalia chebula*), cutch (*Acacia catechu*), gallnut (*Quercus infectoria*), and pomegranate rind and leaf are a few plants that contain tannin dyes (*Punica granatum*) [11].

Dihydropyrans

Flavones and dihydropyran have nearly identical chemical structures. The two most important dyes in this category are hematein and hematoxylin. These vital natural dyes are used on cotton, wool, and silk to create rich hues. Brazil wood (*Caesalpina echennata*), logwood (*Haematoxylum campechianum*), maclurin, and sappan-wood (*Biancaea sappan*) are typical examples (*Chlorfora tinctoria*) [3, 11, 12]. Fig. (**7**) shows the chemical structure of dihydropyran.

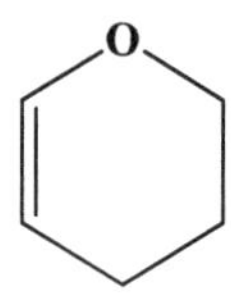

Fig. (7). Chemical structure of dihydropyran.

Carotenoids

Brightly colored natural organic pigments called carotenoids, also known as tetraterpenoids, are present in the chromoplast and chloroplast of almost every plant family, as well as certain other photosynthetic species. Carotenoids can only be synthesized by plants, fungi, and prokaryotes. Long conjugated double bonds are what give the carotenoids their color. They produce the colors yellow, orange, and red by absorbing light in the 400–500 nm range of the spectrum. Among the plants that provide carotenoids are *Bixa orellana*, *Crocus sativus*, *Curcuma longa*, *Nyctanthes arbor-tristis*, and *Cedrela toona* [3]. Many of the bright red, orange, and yellow colors seen in fruits, vegetables, and flower petals are caused by lipid-soluble pigments called carotenoids. The categorization of carotenoids into carotenes and xanthophylls is based on structural distinctions. The components of carotenes are carbon and hydrogen, as opposed to the three components of xanthophylls, which are carbon, hydrogen, and oxygen. Long conjugated dual bonds are what give carotenoids their color. They are in charge of absorbing light in the 400–500 nm spectrum, which produces the colors yellow, orange, and red. Some carotenoid dyes can be found in marigold (*Tagetes erecta*), saffron (*Crocus sativus*), turmeric (*Curcuma longa*), annatto (*Bixa orellana*), and night-blooming jasmine (*Nyctanthes arbor-tristis*) [11, 12]. Fig. (**8**) depicts the chemical structure of carotenoid.

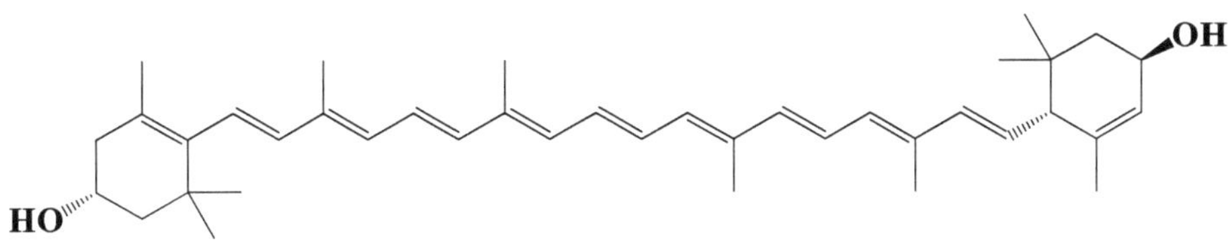

Fig. (8). Chemical structure of carotenoid.

Betalains

A class of water-soluble nitrogen-containing plant pigments belonging to the Caryophyllales order, betalains are comparable to violet betacyanins and yellow betaxanthins. Common natural sources of the betalain class of colorants are *Opuntia lasiacantha* and *Beta vulgaris*, also known as beetroot [3]. The chemical structure of betalain is shown in Fig. (**9**).

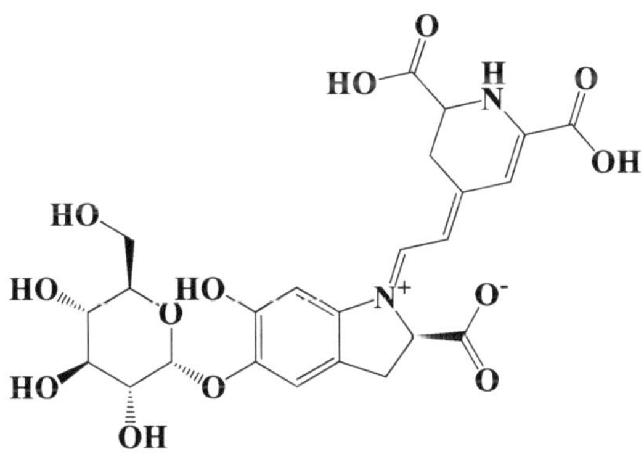

Fig. (9). Chemical structure of betalain.

Based on Application Methods

Natural dyes can be divided into groups according to how they are applied. This category, which includes mordant dyes, direct dyes, acid dyes, basic dyes, vat dyes, and disperse dyes, is frequently linked to the kind of textile material being dyed. Among these, disperse dyes are primarily employed on polyester fabrics, vat dyes are placed on cotton denim, and direct dyes are frequently used for cotton fabric dyeing [11].

Mordant Dye

By adding a mordant, a chemical that improves the interaction between dye and fiber, dyes and colorants that would ordinarily have little to no affinity for a material can be linked to it. All dyes that have the ability to form a complex with the metal mordant are now included in the traditional definition of mordant dyes. The majority of these dyes produce varying hues and tones when mixed with various mordants. [3]. The most natural dyes need mordants to produce long-lasting colors. Mordant dyes must be applied with a mordant since they are not attractive to the textile materials that need to be colored. Complexes are created when these colorants mix with the mordant. Metallic salts (such as aluminium, zinc, copper, tin, cobalt, and potassium), bio-mordants (such as madder, myrobalan, berries, kermes, cochineal, and tannic acid), and oil-based mordants that form complexes with metallic salts are the three different forms of mordants [11, 13].

Vat dyes

Water-insoluble vat dyes, of which indigo and woad are the most notable examples, can undergo reducing conditions to transform into a "leuco" form that dissolves in alkaline aqueous solutions and penetrates the fibers to be colored. These pigment aggregates are trapped in the fiber after being oxidised to their insoluble form by air exposure [14]. The technique of vat dyeing is carried out in a bucket or vat. Although they can be reduced to a soluble colorless (leuco) form that has an affinity for the fiber or textile to be dyed, they are insoluble in their colored state. When the vat dyes are re-oxidized, their original hue is retained, and they are once more transformed into an "insoluble form." Indigo, woad, and tyrian purple are the only three natural dyes that are considered vat dyes [3].

Water does not dissolve certain types of colors. Their most prevalent application is in cellulosic fiber, especially cotton. Colored fibers with excellent color fastness, washability, and chlorine bleachability are produced by vat dyes. Because vat dyes are water-insoluble, they must first be transformed into water-soluble leuco form via a widely used procedure known as the reduction process before being applied to textile materials. The well-known natural vat dyes are woad, indigo, and tyrian purple [11, 15].

Direct Dyes

Unlike vat or mordant dyes, which are applied directly to the fiber, direct dyes are typically less wash- and light-fast. Saffron and turmeric (*Curcuma longa*) are two examples of direct dyes (*Crocus sativus*) [14]. Water-soluble organic compounds known as "direct dyes" can be applied directly to cellulosic fibers like cotton because of their affinity and direct absorption. Direct dyes produce vibrant colors and are simple to apply. Nevertheless, their wash fastness is low because of the nature of the chemical interaction, though this can be enhanced with specialised post-treatment. Safflower, pomegranate, harda, annatto, and turmeric are a few well-known types of direct natural colors [3].

For direct dyes to be attractive to cellulosic fibers, neither the fiber nor the dye needs to undergo any sort of treatment. These dyes dissolve in water and are anionic. Direct dyes have a reputation for being easy to use and providing a wide range of colors and tones, but their washing fastness is lower. Pomegranate rind, annatto, harda, and turmeric are a few examples of direct natural colors [11].

Acidic Dyes

Another kind of direct dye for polyamide fibers, such as wool, silk, and nylon, is acid dye. These dyes, which include either carboxylic or sulphonic acid groups in their molecules, are applied in acidic environments. Saffron is one natural color that has been categorized as an acid dye. There are two carboxylic acid groups in this color [3]. For polyamide fibers like wool, silk, and nylon, acid dyes are an additional kind of direct dye. These dyes are used in acidic environments and contain either carboxylic or sulphonic acid groups in their molecules. Saffron, at least one natural dye, is categorized as an acid dye. Two carboxylic acid groups are present in this dye [11, 15].

Basic Dyes

Catalytic dyes are another name for basic dyes. The carboxyl group of wool and silk fibers forms an electrovalent link with the colorful cations that these dyes produce upon ionization. These colors are used in conditions that range from neutral to slightly acidic. Berberine is categorized as a basic dye. This dye has poor light fastness because of its non-localized positive charge that echoes inside its structure [3]. Basic dyes are categorized as cationic dyes because they generate a colored, positively charged ion in an aqueous medium. Wool and silk's functional group (-COOH) forms an electro-covalent connection with the cationic ions. The dyeing parameter (pH) should be neutral to slightly acidic when using these dyes. These dyes have poor light-fastness characteristics when applied to silk and wool. One example of a natural basic dye is berberine [11, 15].

Disperse Dyes

Polyester and acetate fibers can be dyed with disperse dyes, which are water-insoluble colors. Compared to the era of natural dyeing, the disperse dyeing concept is relatively new. Nonetheless, it is believed that some natural dyes, including lawsone, juglone, lapachol, and shikonin, can be categorized as disperse dyes due to their structural similarities and solubility properties [3]. Disperse dyes have a compact, non-ionic structure and are insoluble in water. These dyes have a lower molecular weight, weak water solubility, and no strong solubilizing groups. It is mainly used to dye hydrophobic textiles like acrylic, acetate, and polyester. They can also be used to dye silk and wool [11, 15].

Natural Dye Chemistry and Application

Chemistry of Natural Dyes

Diverse auxochromic and chromophoric groups are found in different natural colorants. The chemistry of the dyes can be described in terms of their chromophoric groups, depending on whether a specific group is present in the dye molecule. The various chromophoric groups and dye structures are described.

Quinoid-based Structure

The quinoid-based dye structure can be overviewed as three chemical structures: (a) benzoquinone, (b) naphthoquinone, and (c) anthraquinone. The natural colorant carthamine belongs to the benzoquinone group, whereas juglone and lawsone have the naphthoquinone structure. Alizarine dye possesses an anthraquinone structure. Fig. (**10**) shows the chemical structure of carthamine (a), juglone (b), and lawsone (c).

Benzoquinone Dyes

The л electron system is tiny in this dye structure, and the dye has another unsaturated group conjugated to it. Safflower contains the red colorant carthamine (Natural Red 26). The subtropical plant known as safflower (*Carthamus tinctorius*) is grown throughout North and South America, Europe, China, and India. In the dyeing process, cold water is used to extract the water-soluble yellow dye (safflor yellow), and a diluted sodium carbonate solution is used to extract the red safflorcamin. Wool, silk, and cotton can be dyed with the extracted solution once it has been neutralised [16].

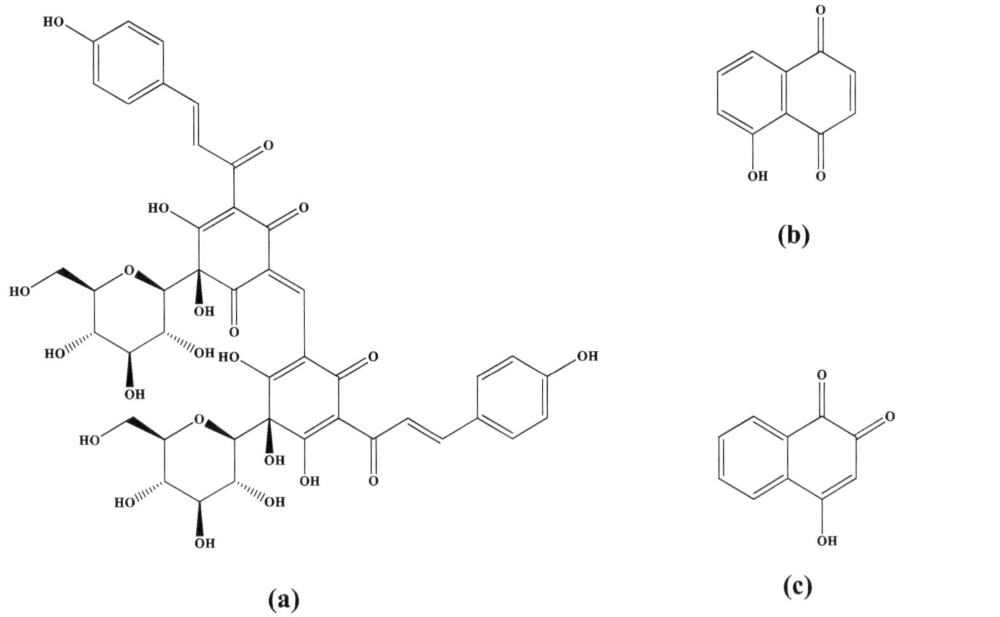

Fig. (10). Chemical structure of carthamine (a), juglone (b), and lawsone (c).

Naphthoquinone Dyes

This group includes natural dyes utilizing Lawsone and Juglone. Lawsone is taken from the Henna plant, which also has luteolin, a flavonoid colorant, in its leaves. It is grown in nations including Australia, Africa, and India. Hennosides A, B, and C are glycosidic forms of naphthoquinone. Because the dye molecules' structures resemble those of disperse dyes, henna can be used to dye cotton, polyester, polyamide, and cellulose triacetate. A natural dye with a naphthoquinone composition is juglone. Different parts of nut trees are used to obtain the dyestuff. Trees and plants contain juglone in the glycoside form. Juglone-dyed wool exhibits good insect and moth resistance. The qualities of fastness are further improved by mordanting. A brown hue is produced when textile fabrics are dyed with aqueous walnut extract. Juglone can be used to color a variety of textile fibers, including wool, silk, nylon, and polyester [16].

Anthraquinones

It has the largest collection of anthraquinone dyes. The plant's roots are used to extract rhubarb (CI Natural Yellow 23). Emodin, chrysophenol, aloe emodin, and pyscion are all present in the extracted dye. Wool fiber is dyed with rhubarb extract. After being mordanted with alum, it takes on a yellow to orange hue. The mordanting process increases the dyed materials' light fastness. Alizarin, pseudo-purpurin, and purpurin are natural dyes that are anthraquinone-structured and belong to the *Rubiaceae* family of plants. The plant's roots are used to make the dye. In Europe, Asia, and Northern and Southern America, madder is grown as a raw material for red colorant. Madder (C.I. Natural Red 8) is a natural dye. The plant's dried roots are used to extract the color. Two to three percent of di- and tri-hydroxyl anthraquinone glucosides are found in the plant's roots [17].

Carotenoids

Plants and animals have red, yellow, and orange pigments called carotenoids. Its structure is polyisoprenoid, with a number of conjugated bonds positioned in the centre. Carotenoids give many fruits and vegetables their vibrant colors. As polyisoprenoid structures, carotenoids have conjugated double bonds that function as chromophores and give them their distinctive absorption spectra. There are two components of carotenoids: A) hydrocarbon carotenoids, and B) xanthophylls, containing an oxygen moiety. Numerous carotenoid structures were produced by structural alterations brought about by hydrogenation, double bond migration, isomerization, and chain lengthening and shortening. Carotenoids possess strong resistance to UV light, and β-carotene is a common structure found in natural colorants. [18].

Pyron Dyes

Flavonoids and anthocyanins with a structure are found in pyron dyes. Glycosidic linkages hold the pyran structure to a variety of sugars. Flavonols, flavones, anthocyanidins, isoflavones, flavon-3,4-diols, and coumarins are the different types of flavonoids. Vegetable dyes are made from flavonols and yellow flavones. Yellow quercetin is a valuable and well-liked flavonoid with a number of biological effects [12].

Anthocyanins

The flavonoid class of natural pigments includes anthocyanins, which are primarily employed as natural colorants. Grape wine, sweet and sour cherries, red cabbage, hibiscus, and several kinds of oranges are among the fruits and vegetables that contain them. Anthocyanins, which generate the colors red, pink, violet, and orange, come in over 500 different kinds. Cyanidin, delphinidin, pelargonidin, malvidin, peonidin, and petunidin are a few significant anthocyanins (Fig. **11**). In addition to anthocyanins, many plants also contain quercetin and chlorophylls, and the combined color is a combination of these substances [19].

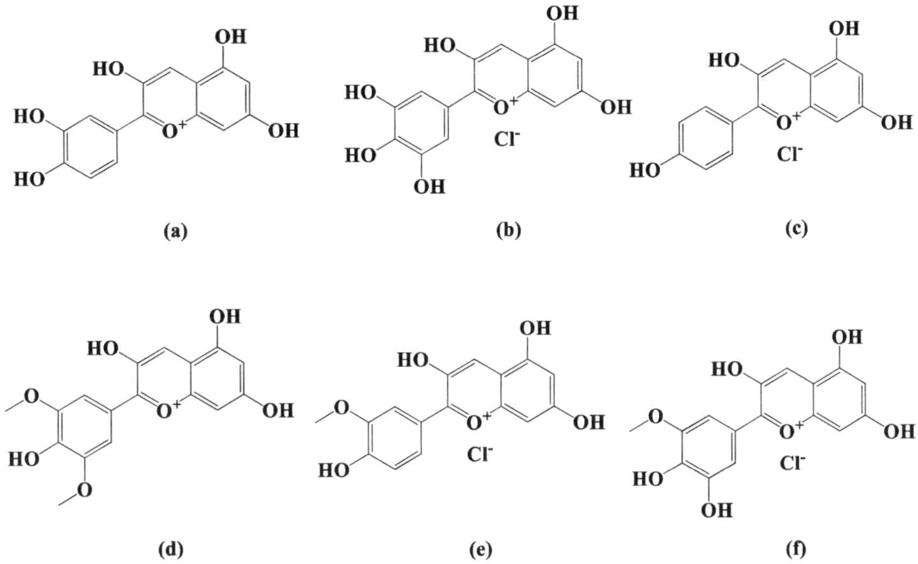

Fig. (11). Chemical structure of Cyanidin (a), Delphinidin (b), Pelargonidin (c), Malvidin (d), Peonidin (e), Petunidin (f).

Dyes from Lichens and Mushrooms

Throughout antiquity until the start of the Middle Ages, mollusks and shellfish were the primary source of violet and purple pigments. Originally derived from mollusks, the color was known as royal purple or Tyrian purple. Natural dyes can be found in lichens and mushrooms, which yield purple and violet hues. Coastal regions are home to lichens, which are simpler to gather. The drawback of lichens is their weak light fastness; however, the dyeing techniques are simple. Lichen dyeing is therefore restricted to low-quality textiles. Textiles can also be dyed using fungi. The fungus *Echinodontium tinctorium* is used to produce red in America and India. Wool was dyed in France and Italy using fungi derived from Polyporales. Benzoquinone derivatives, particularly terphenylquinone, are the colorants found in lichens and fungi. Sarcodon, Phellodon, Hydnellum, and Thelephora are among the chemicals found in some of these species. The coloring agents that give lichens their color are litmus and orchid.

The pre-compounds of orchil and litmus, which are created by successive enzymatic hydrolysis, decarboxylation, and oxidizing reactions, respectively, give the lichens their color. Lecanoric acid, atranorin, and gyrophoric acid are three pre-compounds that contribute to the synthesis of litmus and orchil. In the past, lichens were kept in ammonia-infused water for a few days in order to extract colorants from them. The reaction was caused by enzymatic hydrolysis, which hydrolyzes and decarboxylates non-colored substances like lecanoric acid to produce orcinol. After oxidation, orcinol produces litmus or purple orceins. The pH of the solution affects the color of litmus and orchil. Dyestuff produces a bluish violet anion in basic pH and a red cation in acidic pH. When used to dye wool, lichens from the species *Parmelia, Xanthoria parietina, Ochrolechia tartarea*, and *Lasallia pustulata* can give yellowish, brownish, and reddish-brown hues. The wool is dyed by boiling it in lichen solution, either with or without mordanted wool, while ammonia is present. Terphenyl quinone chemicals are the primary colorants that give mushrooms their blue hue and are found in species Sarcodon, Phellodon, and Hydnellum. They are derivatives of benzoquinone. The mushrooms of the Cortinarius species include vibrant shades of violet, olive green, scarlet, and brown. They are derivatives of anthraquinones [20, 21].

Tannins

Tannins are polymeric polyphenols with a comparatively high molecular weight and a characteristic aromatic ring structure with hydroxyl components. Plants contain two distinct types of tannins: proanthocyanidins and hydrolysable tannins (condensed tannins). Plant cells contain tannins, which are mostly found in the tissues of the epidermis. Wood, leaves, buds, stems, flowers, and roots all contain tannins. Many plants have a concentration of hydrolysable tannins in their roots. Many types of tannins are found in plants. Gallic acid and ellagic acid are the two groups of hydrolysable tannins known as gallotannins and ellagitannins (Fig. **12**). Pentagalloyl glucose is the most common gallotannin. Hexahydroxydiphenic acid esters are known as ellagitannins. Certain hydrolysable tannins include both gallic acid and hexahydroxydiphenic acid. Polymers comprising 15-carbon polyhydroxyflavan-3-ol monomer units, such as (−) epicatechin or (+) catechin, are known as condensed tannins (Fig. **13**). Although there are some documented pathways for biosynthesis, the biosynthesis and polymerization of tannins are challenging due to their complicated chemical composition. Shikimic acid is the starting point for the production of hydrolysable tannins. Gallic acid is created when 3-dehydroshikimic acid is directly aromatized. This acid then undergoes esterification to generate a polyol. Condensed tannins are biosynthesized in two ways: (a) by phenylpropanoid and (b) by polyketide. In the biosynthesis of flavonoids, the polyketide pathway uses malonyl moieties to create aromatic rings. The aromatic amino acid L-phenylalanine is taken up via the phenylpropanoid pathway and non-oxidatively deaminated by phenylalanine ammonia-lyase to E-cinnamate. [12].

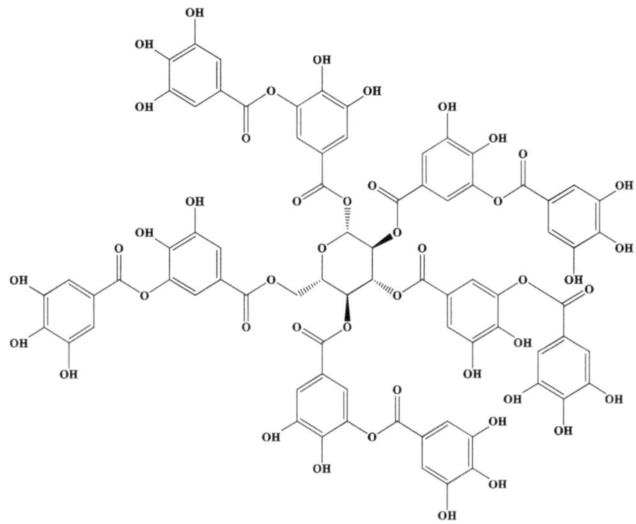

(a)

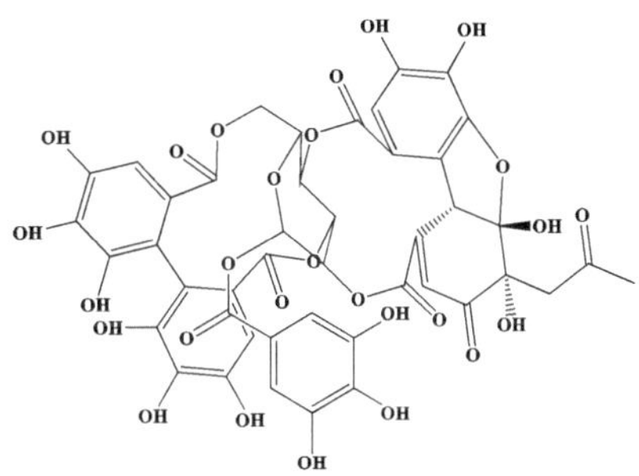

(b)

Fig. (12). Chemical structures of gallotannins (a) and ellagitannins (b).

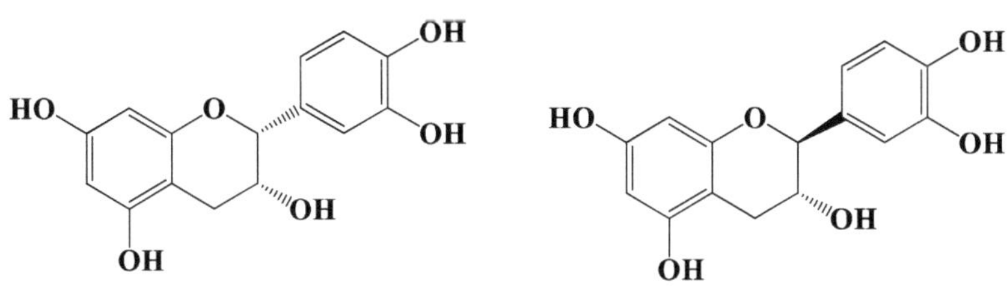

(a) **(b)**

Fig. (13). Chemical structures of epicatechin (a) and catechin (b).

CONCLUSION

Natural dyes are multi-source, vibrant gifts from the natural world. Minerals, microbes, plants, and animals can provide a variety of lovely and practical dyes. Mineral dyes are stable, plant-based dyes are environmentally benign, animal-based dyes are vibrant and long-lasting, and microbial dyes are a developing field of study. Understanding the chemical makeup of natural dyes, such as flavonoids or anthraquinones, helps us understand why they have varying hues and how effectively they adhere to textiles. People can utilize them more efficiently if they know how to apply them, regardless of whether they require mordants or not. Natural colors are increasingly being used in food, cosmetics, fabrics, and even medicine as people adopt more environmentally friendly lives. This chapter demonstrates how natural dyes are not only a part of our past but also play a significant role in sustainable color in the future.

ACKNOWLEDGMENTS

The authors wish to recognize the application of AI tools, like Grammarly, during the writing process to enhance the clarity and language of the manuscript. The authors thoroughly reviewed and modified the content as necessary and accept full responsibility for the content of the chapters. Additionally, they want to emphasize that AI was not utilized in our chapters to create or modify images, aid in data analysis, or contribute to our research.

REFERENCES

[1] Saxena S, Raja A. Natural dyes: sources, chemistry, application and sustainability issues. Roadmap to sustainable textiles and clothing: eco-friendly raw materials, technologies, and processing methods. Springer 2014; pp. 37-80.
[http://dx.doi.org/10.1007/978-981-287-065-0_2]

[2] Yadav S, Tiwari KS, Gupta C, Tiwari MK, Khan A, Sonkar SP. A brief review on natural dyes, pigments: Recent advances and future perspectives. Results in Chemistry 2023; 5: 100733.
[http://dx.doi.org/10.1016/j.rechem.2022.100733]

[3] Yusuf M, Shabbir M, Mohammad F. Natural colorants: Historical, processing and sustainable prospects. Nat Prod Bioprospect 2017; 7(1): 123-45.
[http://dx.doi.org/10.1007/s13659-017-0119-9] [PMID: 28093670]

[4] Lara L, Cabral I, Cunha J. Ecological approaches to textile dyeing: a review. Sustainability (Basel) 2022; 14(14): 8353.
[http://dx.doi.org/10.3390/su14148353]

[5] Alegbe EO, Uthman TO. A review of history, properties, classification, applications and challenges of natural and synthetic dyes. Heliyon 2024; 10(13): e33646.
[http://dx.doi.org/10.1016/j.heliyon.2024.e33646] [PMID: 39040261]

[6] Salauddin SkM, Mia R, Haque MA, Shamim AM.. Review on extraction and application of natural dyes. Textile & Leather Review 2021; 4(4): 218-33.

[7] Borges ME, Tejera RL, Díaz L, Esparza P, Ibáñez E. Natural dyes extraction from cochineal (Dactylopius coccus). New extraction methods. Food Chem 2012; 132(4): 1855-60.

[http://dx.doi.org/10.1016/j.foodchem.2011.12.018] [PMID: 23442630]

[8] Jain M. Ecological approach to reduce carbon footprint of textile industry. International Journal of Applied Home Science 2017; 4(7/8): 623-33.

[9] Li N, Wang Q, Zhou J, Li S, Liu J, Chen H. Insight into the progress on natural dyes: sources, structural features, health effects, challenges, and potential. Molecules 2022; 27(10): 3291. [http://dx.doi.org/10.3390/molecules27103291] [PMID: 35630767]

[10] Vankar PS. Chemistry of natural dyes. Resonance 2000; 5(10): 73-80. [http://dx.doi.org/10.1007/BF02836844]

[11] Repon MR, Dev B, Rahman MA, *et al.* Textile dyeing using natural mordants and dyes: a review. Environ Chem Lett 2024; 22(3): 1473-520. [http://dx.doi.org/10.1007/s10311-024-01716-4]

[12] Gupta VK. Fundamentals of Natural Dyes and Its Application on textile substrates. In: Chemistry and Technology of Natural and Synthetic Dyes and Pigments. IntechOpen; 2020.

[13] Prabhu K, Bhute AS. Plant based natural dyes and mordants: A Review. J Nat Prod Plant Resour 2012; 2(6): 649-64.

[14] Ferreira ESB, Hulme AN, McNab H, Quye A. The natural constituents of historical textile dyes. Chem Soc Rev 2004; 33(6): 329-36. [http://dx.doi.org/10.1039/b305697j] [PMID: 15280965]

[15] Benkhaya, benkhaya, El Harfi, S., & El Harfi, A. (2017). Classifications, properties and applications of textile dyes: A review. Applied Journal of Environmental Engineering Science, 3(3), Appl.00000J. Envir. Eng. Sci. 3 N°3(2017) 311–320. [http://dx.doi.org/10.48422/IMIST.PRSM/ajees-v3i3.9681]

[16] Bechtold, T. and R. Mussak, Natural colorants–quinoid, naphthoquinoid and anthraquinoid dyes. Handbook of natural colorants. A John Willey and Sons Publication 2009; 23(1): p. 151-182. [http://dx.doi.org/10.1002/9780470744970.ch10]

[17] Diaz-Muñoz G, Miranda IL, Sartori SK, Rezende DC, Diaz MA. Anthraquinones: an overview. Studies in natural products chemistry, 2018. 58: p. 313-338.

[18] Chandrika, U.G., Carotenoid dyes-properties. Handbook of natural colorants, 2009: p. 221-236. [http://dx.doi.org/10.1002/9780470744970.ch13]

[19] Amogne NY, Ayele DW, Tsigie YA. Recent advances in anthocyanin dyes extracted from plants for dye sensitized solar cell. Mater Renew Sustain Energy 2020; 9(4): 23. [http://dx.doi.org/10.1007/s40243-020-00183-5]

[20] Räisänen, R., Natural colorants from lichens and mushrooms. Handbook of Natural Colorants, 2023: p. 317-331.

[21] Räisänen R. Dyes from lichens and mushrooms Handbook of Natural Colorants. 2009; pp. 183-200. [http://dx.doi.org/10.1002/9780470744970.ch11]

CHAPTER 3

Natural Dyes *vs.* Synthetic Dyes: Pros, Cons, and Sustainability

Naheed Akhtar[1,*], **Syed G. Musharraf**[1,2,3] and **Vinitha M. Thadhani**[4]

[1] *H.E.J. Research Institute of Chemistry, International Center for Chemical and Biological Sciences, University of Karachi, Karachi, Pakistan*

[2] *Dr. Panjwani Center for Molecular Medicine and Drug Research, International Center for Chemical and Biological Sciences, University of Karachi, Karachi, Pakistan*

[3] *School of Chemistry and Pharmaceutical Sciences, Guangxi Normal University, Guilin, Guangxi, China*

[4] *Ceylon Ayurveda Pvt. Ltd, Havelock Road, Colombo, Sri Lanka*

Abstract: This chapter examines the characteristics, benefits, drawbacks, and effects on the environment of natural and synthetic dyes. Natural dyes are derived from minerals, plants, and animals, whereas synthetic dyes are created in labs using chemicals. This chapter describes how both kinds of dyes function, how to apply them, and how their color strength, fastness, consistency, and simplicity of use vary. It also draws attention to the effect of both dyes on the environment and carbon footprint. Knowing their advantages and disadvantages can help us choose more effective and sustainable dying techniques.

Keywords: Color fastness, C footprinting, Greenhouse gases, Natural fibers, Synthetic dyes.

INTRODUCTION

Textiles, paper, food, and other objects can be colored with dyes. Natural dyes, which are derived from nature, and synthetic dyes, which humans create through chemical processes, are the two primary categories of dyes. Each has special qualities and applications. Minerals, microbes, plants, and animals can all be used to make natural dyes. Although they are renewable, biodegradable, and less damaging to the environment, they might not provide vibrant, long-lasting color without proper care. Because of their wide range, vibrant hues, and reliable performance, synthetic dyes are produced in laboratories and are used extensively

* **Corresponding author Naheed Akhtar:** H.E.J. Research Institute of Chemistry, International Center for Chemical and Biological Sciences, University of Karachi, Karachi, Pakistan; E-mail: naheedakhtar41@yahoo.com

in the textile industry. But they frequently use hazardous substances that can contaminate the environment. In this chapter, natural and synthetic dyes are compared according to their manufacturing processes, ease of application, consistency, and environmental impact. It assists readers in weighing the benefits and drawbacks of each and determining which is best suited for specific uses.

Before the first synthetic dye was discovered in 1856, the only dyes that humans could use to color fabrics were those made from natural materials, including plant leaves, roots, bark, insect secretions, and minerals. Rapid advancements in synthetic chemistry, aided by the industrialization of textile production, produced a variety of synthetic dyes in a wide range of hues and colors that eventually drove natural dyes to extinction, in addition to creating synthetic substitutes for widely used natural dyes. However, in the final decades of the 20th century, consumer interest in natural dyes was once again rekindled by environmental concerns surrounding the manufacture and use of synthetic dyes. There is already a niche market for fabrics dyed with natural dyes, which are preferred by people who care about the environment. However, because of specific technical and sustainability issues with their production and application, such as their limited and non-reproducible shades, unsuitability for machine use, and lack of readily available standard forms, the overall share of natural dyes in the textile industry is only about 1%. Although natural dyes are sustainable in and of themselves because they are biodegradable and renewable, the textile industry's enormous demand cannot be met by them due to the preferential use of land for food and feed. Additionally, overusing natural resources to make colors might imperil endangered species and cause deforestation. Because of these factors, the Global Organic Textiles Standard (GOTS) forbids the use of natural colors derived from endangered species while allowing the use of safe synthetic dyes. Given their enormous environmental benefits, numerous research projects have been conducted worldwide to address the drawbacks of natural dyes [1].

Fundamentals of Natural Dyes and Synthetic Dyes

The chemical makeup of various materials determines how intensely they color, and dye is a substance that can bind to other substances. The earliest coloring agents were from plants (foh, indigo, *etc.*) and animals (*e.g.*, cochineal). Nowadays, dyes have been produced, especially those with basic and acidic characteristics, and they are employed in food and clothes. Natural dyes and synthetic dyes, which are categorized into several varieties such as azo, carmine, and saffron, are the two categories of dyes. In practically every aspect of our everyday lives, dyes are used to color clothing, food, cosmetics, paper, and other materials. For almost 4,000 years, textile dyes have been made from natural plant, animal, and mineral pigments [2].

Natural Dyes

In essence, dye is a color that is typically applied to various materials to stain them. Around 2600 BC, the first natural dye was discovered. Natural pigments were first combined with water or oil as a foundation to create colors. Plants (like indigo and saffron), insects (like cochineal beetles and lac scale insects), animals (like particular species of mollusks or shellfish), microbes (like pseudomonas, bacillus, and rhodococcus), and minerals (like ferrous sulphate, ochre, and clay) can all be used to make natural dyes without the need for chemicals [3].

Nearly every kind of natural fabric may be dyed using natural dyes. They can also be used to dye certain synthetic textiles, according to recent studies. In addition to being utilized in textiles, natural dyes are also employed in the coloring of food, medications, toys, handicrafts, and leather. Many plants that produce dye are also used as medicines in a variety of traditional medical systems [1]. Natural colorants have the following advantages as compared to synthetic colorants, but natural dyes have some disadvantages too.

Natural dyes are non-toxic, non-polluting, less harmful to health, and non-carcinogenic. Compared to synthetic dyes, the majority of natural dyes are regarded as environmentally beneficial since they are generated from renewable resources (derived from non-renewable petroleum resources and synthesized in an intermediate route involving many chemical hazards). They produce a calming shade, have colors that harmonise, and are delicate, soft, and subtle.

In contrast to the non-renewable fundamental raw materials used in synthetic dyes, natural colors derived from plants are typically agro-renewable or vegetable-based goods that are also biodegradable. Numerous benefits of natural dyes include their non-toxicity, environmental friendliness, eye-pleasing hue, unique scent, and freshness of shade. Simple extraction and purification, no production of wastewater, Renewable resources, mild dyeing conditions, and extremely high sustainability [4]. Numerous natural dyes and finishing chemicals have antibacterial and UV-protective properties. Thus, natural textiles that are dyed with appropriate natural colorants and polished with particular natural finishing agents can offer defense against UV radiation, bacteria, and even mosquito bites. As with Ayurvedic medicine, natural dyes such as myrobalan, turmeric, madder/manjistha (MJ), Arjuna, safflower, *etc.*, have therapeutic benefits [5]. With a mix-and-match technique, natural dyestuffs can also create a vast array of hues. Numerous new hues can be produced with a very slight change in the dyeing method, the use of different mordants with the same dye, or the application of varied mordant concentrations on the same dye [6].

Currently, the sector needs around 3 million tons of dyestuff. Given this, using natural colors in conventional textile processes is extremely difficult. The main purposes of agricultural land are to raise animals and feed the world's growing population. Additionally, the sustainability of natural dyes is a significant concern, and biodiversity shouldn't be sacrificed for dye extraction [1]. The use of natural dyes has a number of restrictions and difficulties. Poor color reproducibility, poor or uneven composition, average washing fastness, and limited availability in various regions are some of the drawbacks of natural dyes that are seriously impeding their resurgence. Furthermore, there is no acknowledged standard dyeing technique for natural dyes. The type of mordant used in the dyeing process—some of which are not environmentally friendly—determines the final coloration. Fabrics made of cotton, linen, wool, silk, nylon, and polyester are dyed using natural dyes [4].

Compared to the time needed to apply synthetic dyes to the same textiles, the extraction, purification, and actual dying process that involves mordanting takes longer for natural dyes. Steps in the dye extraction process take more effort and preparation. Although natural dyes are more environmentally friendly than synthetic dyes, most natural dyes have poor exhaustion on textile materials, even when mordants are used. This results in a significant amount of colorant component remaining in the dye bath after dyeing.

The spectrum of shades that may be obtained from natural dyes and pigments is limited, making it challenging to achieve all desired tones. Although there are multiple sources for reddish and greenish dyes, there is only one significant source for the blue natural dye, which comes from natural indigo, out of the three primary hues that are necessary: red, green, and blue. Only a few natural dyes can be put together in a binary or ternary mixture due to the differences in their chemistry and application methods; these dyes are obviously incompatible when creating compound hues [6].

These agro-products vary from crop season to crop season in terms of location, species maturity duration, and continuous shadow because of the different proportions of ingredients. However, even with the extract of different concentrations of colored extract of such selective natural dyes, the reproducibility of shades can be better guaranteed if the dye extract can be diluted to bring it to the same concentration level of colored solution after extraction by UV spectral analysis [5].

Synthetic Dyes

Following the development of synthetic dyes in the latter part of the nineteenth century, the use of natural colors began to decrease. Due to their easy availability in ready-to-apply form, ease of application, consistency of shades, and superior fastness properties, synthetic dyes have nearly completely replaced natural dyes as a result of coordinated research efforts in the field and the quick industrialization of textile production [1]. Perkins William made the first synthetic organic dye discovery in 1856 with the discovery of purple, which is an aromatic chemical that produces a wide range of vibrant hues [7]. We were looking for a method to create chitin, a medication used to treat malaria. Therefore, Perkins Henry created a new substance to produce the dye in his effort to manufacture chitin. In order to create the synthetic organic dye, Woufflé used the natural color indigo with nitric acid in 1871 to create picric acid. Numerous additional chemical dyes have been added to the list of dyes since then. At the start of the twentieth century, synthetic dyes nearly entirely supplanted natural colors. In contrast to today's synthetic colors made from aromatic organic compounds, which still have tar as their primary source, many natural dyes have been known since ancient times and were derived from plant and animal sources (Coal Tar). Tar from Coal [2]. Numerous sectors make substantial use of synthetic dyes, with the textile manufacturing sector being one of the biggest users [3].

Classification of Synthetic Dyes

Chemical substances are used in the artificial synthesis of synthetic colors. Depending on the type of created fiber, synthetic dyes are categorized into three types. These include colors for cellulose, protein, and synthetic fibers. The classification of synthetic dyes is shown in Fig. (**1**).

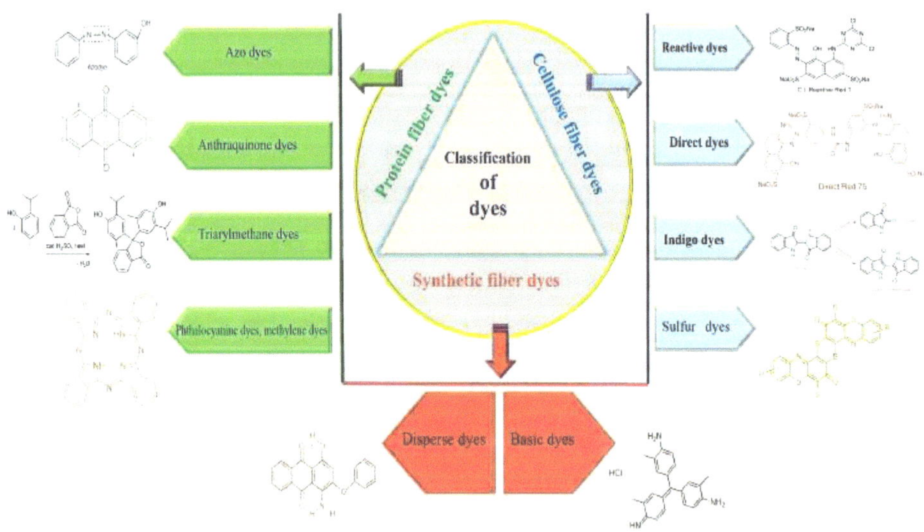

Fig. (1). Classification of synthetic dyes.

By providing an exceptional variety of hues, reliable dyeing outcomes, and expedited production processes, synthetic dyes helped to establish the textile and fashion sectors. The wide range of hues that synthetic dyes provide is one of their main benefits. Synthetic dyes' chemical makeup can be adjusted to produce a wide range of colors, something that is challenging to accomplish with just natural dyes. This multiplicity of colors has impacted the aesthetics of contemporary textile and apparel design. With shorter dyeing durations and improved color absorption qualities, synthetic dyes allowed for the unprecedented mass production of textiles, lowering costs and increasing accessibility for a wider variety of customers [8]. The industries of textile dyeing, paper printing, color photography, pharmaceuticals, food, cosmetics, and leather all make substantial use of synthetic dyes [9]. Because the raw chemicals are widely accessible and the manufacturing procedures are very effective, producing synthetic colors is typically less expensive. Because of their low cost, synthetic dyes are widely utilized in a variety of industries, including printing, textiles, and cosmetics [3, 10].

Because of their chemical makeup and structure, synthetic dyes are not biodegradable, which has a negative impact on the environment. They also tend to be highly stable to common oxidation and reduction processes according to their design, making them extremely challenging to remove from textile industry effluents [3, 11]. Petrochemicals are used to make synthetic dyes, which pollute the environment [8]. Because of their synthetic origin and intricate chemical architectures, which hinder their capacity to biodegrade, synthetic dyes pose a significant risk to the environment and water streams. Humans who use synthetic dyes may experience breathing issues, allergies, and skin irritation. Synthetic dyes have the potential to release harmful chemicals and pollutants into the land, water, and air, which can lead to soil contamination, greenhouse gas emissions, and air and water pollution [8].

Reproducibility and Consistency of Synthetic Dyes and Natural Dye Production

Although natural dyes typically lack the intense brilliance seen with synthetic dyes, they frequently produce distinctive, gentler, and earthy aesthetics. Furthermore, compared to synthetic dyes, which provide exceptional color consistency and stability and retain their brightness even after numerous washings or extended exposure to light, natural dyes fade more quickly in the presence of sunshine or after washing. Because of this feature, synthetic dyes are more suited for uses like outdoor fabrics where color fastness is crucial. The expense of producing natural dyes is high since it takes a long time and involves more human involvement. However, because the raw materials are easily accessible and the manufacturing techniques are very effective, the creation of synthetic dyes is typically more affordable. Because of their low cost, synthetic dyes are widely utilized in many different industries, including [8]. Because of their chemical stability, resistance to deterioration, capacity to produce consistent hue through repeatable procedures, and affordability, synthetic dyes are favored in the industrial sector. Accurate dyeing outcomes utilizing raw materials can be challenging to accomplish because of the complexity of mordanting techniques, raw material variability, and challenges in controlling dyeing conditions. The composition of dyes is influenced by seasonal, environmental, and geographic factors; minor changes in temperature, concentration, or treatment time can produce uneven coloring [12].

Ease of Application of Synthetic Dyes and Natural Dyes

Understanding the characteristics of both natural and manufactured dyes is crucial for comprehending their uses, behavior, and interactions with different substrates. There are several qualities that both natural and manufactured dyes have, such as color, solubility, stability, and material affinity. Selecting the appropriate dye for a given application requires an understanding of these characteristics in order to meet the requirements for safety, durability, and desired color. Chemical stability, color fastness, solubility, and substrate affinity are a few of the characteristics of dyes [8].

Color Fastness

The capacity of dyes to maintain their color in the face of environmental influences, including light, washing, and chemical exposure, is known as color fastness. It is an essential characteristic to guarantee the durability of dyed fabrics. Usually, standardized testing techniques that mimic actual conditions are used to evaluate color fastness. Various forms of color fastness are used to assess specific dyeing properties. These consist of the following: chemical, migration, sublimation, light, rubbing, sweating, and wash fastness.

In practical applications, wash fastness ensures longevity by measuring a dye's resistance to fading or bleeding during washing. Uneven coloration or color loss may result from the removal of poorly fixed or unstable dyes from fabrics subjected to water, detergents, and mechanical agitation during washing. For textiles that are frequently cleaned in water, wash fastness is particularly crucial. To evaluate it, dyed textiles are repeatedly washed, and color changes are monitored. The ability of dyes to hold their color through numerous washings is known as wash fastness. The degree of color shift is described by ratings, ranging from excellent to poor.

The ability of fast dyes to withstand fading in the presence of light and other environmental conditions is well-known. They are frequently utilized in outdoor applications, including fabrics for automobiles. Certain dyes are made specifically to be light-sensitive, changing color in response to light. Applications for this include sun-sensitive textiles. A variety of wavelengths found in light, particularly UV radiation, can degrade dye molecules. Poor light-fastness dyes can fade, alter color, or even break down the dye molecules. The ability of a dye to hold its color in the presence of light, especially sunshine, is measured by its light fastness. It describes how dyes can withstand fading in the presence of artificial or solar light, guaranteeing a long-lasting tint. In order to test for light fastness, dyed samples are exposed to regulated light sources for specific amounts of time. The color of the exposed samples is then compared to reference samples that were not exposed. Based on the amount of fading seen, light fastness is assessed on a scale that typically goes from 1 (poor) to 8 (outstanding). In order to replicate sunshine, it is frequently carried out with tools like a Xenon arc lamp.

Rubbing fastness, often called crocking fastness, is a test that determines how resistant dyes are to color transfer when they come into contact with another surface. For textiles that come into contact with other materials while being worn, it is crucial. The ability of dyes to withstand fading in the presence of human perspiration is measured by their perspiration fastness. Perspiration fastness tests come in a variety of forms to mimic different sweating intensities. Chemical fastness tests evaluate a dye's resistance to color change in the presence of different chemicals, including acids, alkalis, and everyday household items. The ability of dyes to move from one part of a fabric to another is measured by migration fastness. It is crucial for stopping color bleeds in textiles with several colors. Sublimation fastness evaluates whether dyes will undergo color loss or stains if they sublime, or change straight from a solid to a gas, under particular circumstances [8].

Solubility

Dyes need to dissolve in the solvent or water to which they are applied. They can efficiently pierce the material being dyed thanks to this feature. Their structure influences the solubility of dyes in various solvents. Because water-soluble dyes can mix with fibers and form long-lasting color bonds, they are frequently used to dye textiles. One crucial characteristic of dyes is their solubility, which establishes their capacity to dissolve in a specific solvent or medium. The solubility of a dye has a direct impact on how well it penetrates and interacts with the materials it is used on, such as paper, textiles, and other materials.

The ability of a dye to dissolve in water or other solvents to form a dyeing solution is known as dye solubility. Water-soluble dyes can be used to color fabrics, paper, and other materials in aqueous solutions because they dissolve easily in water. They are frequently employed in sectors where water-based dyeing techniques are favored because of their user-friendliness and lower environmental impact. Acetone and ethanol are examples of organic solvents in which solvent-soluble pigments dissolve. They are used in fields like marker inks and specific industrial procedures where water-based solutions are inappropriate. Oil-soluble dyes can be used to color lipsticks, cosmetics, and other oil-based items since they dissolve in fats and oils.

Temperature, pH, ionic strength, and chemical structure are some of the variables that affect a dye's solubility. The solubility of dyes is influenced by their chemical structure, which includes their functional groups and general polarity. Compared to organic solvents, polar dyes are more prone to dissolve in water. While dyes with hydrophilic groups are more likely to dissolve in water, those with hydrophobic groups are more likely to dissolve in nonpolar solvents. Solubility is also affected by temperature. While some dyes behave oppositely, others become more soluble at higher temperatures. Specific temperature ranges are necessary for the best color development of certain dyes. Thermochromic dyes react to temperature changes by changing color. Applications for them include temperature-sensitive labels and mood rings.

Dye solubility is influenced by the pH of the solution. While certain dyes dissolve better in alkaline solutions, others are more soluble in acidic ones. Through ion-dye interactions, the presence of salts or other ions in a solvent also affects how soluble dyes are. A dye's solubility is carefully controlled in several dyeing procedures to get the required color intensity and consistency. The substrate and the intended result determine whether solvent or dyeing medium is best. For instance, because water-soluble dyes work well with natural fibers like cotton and wool, they are frequently employed in textile dyeing [8].

Affinity for Substrates

Differing substrates, including paper, plastics, synthetic fibers (polyester, nylon), and natural fibers (cotton, wool), have different affinities for dyes. To get the desired color output, the dye selection is based on substrate compatibility. The degree to which a dye molecule is drawn to and interacts with the substance it is meant to color is known as its affinity for substrates. Numerous elements affect this feature, such as the chemical makeup of the dyes and substrates, the interaction and intermolecular forces between the dyes and substrates, and the compatibility of size and shape.

Dye compatibility with various substrates is influenced by its molecular structure. Certain fibers or materials are more receptive to dyes with particular functional groups. Substrates' interactions with dyes are influenced by their structure and composition. Because of their varied surface properties and chemical makeup, natural fibers such as cotton, wool, and silk have diverse affinities for various dyes.

Affinity is influenced by substrate surface characteristics as well as the size and structure of the dye molecule. For consistent, vivid coloring, dyes must work well with the substrate and the dyeing procedure. Reactive dyes, for instance, have functional groups that can create covalent connections with particular cellulose fiber sites, resulting in a robust and long-lasting attachment. However, because of their capacity to create ionic connections, acid dyes are frequently used on protein fibers like wool and silk [8].

Chemical Stability of Dyes

Dye needs to be chemically stable in the conditions it is exposed to in order to prevent color changes or deterioration over time. The ability of dyes to resist chemical changes or degradation under a variety of environmental conditions is known as their chemical stability. For dyed materials to maintain their proper color and appearance over time, dyes must be chemically stable. Several variables, including molecular structure, UV light, heat, chemical resistance, moisture and humidity, and pH, affect the chemical stability of dyes.

The chemical composition and structure of dyes have a significant impact on their stability. Chemical processes that result in color change or degradation are less likely to occur in dyes with stable molecular structures. When exposed to ultraviolet (UV) light, certain dyes deteriorate and lose some of their color intensity. Dyes with conjugated systems or protective groups may exhibit higher UV stability. For heat-related applications like textile dying or printing, dyes' ability to withstand high temperatures without experiencing significant degradation is essential. Certain dyes are suitable for applications involving heat because they remain stable at high temperatures.

Colors are usually subjected to a range of chemicals during washing, such as detergents and other cleaning agents that could contain acids. When exposed to these substances, dyes with a high level of chemical resistance maintain their color and structure. Certain dyes are susceptible to fading, color bleeding, and other degradations due to their sensitivity to moisture and humidity. As a result, moisture resistance is especially important for textiles and outdoor applications. Due to their sensitivity to pH changes, dyes undergo color changes or degradation in conditions that are either acidic or alkaline. This trait is particularly helpful in chemical and biological applications. One important aspect affecting the color stability of natural dyes is their pH stability. The pH levels at which natural dyes reach their maximum color intensity vary. Consequently, the dye solution's pH needs to be adjusted for the specific use [8].

C Footprinting of Synthetic Dyes and Natural Dyes

The textile industry is one of the primary causes of environmental contamination. Emissions of greenhouse gases (GHGs) and carbon dioxide (CO_2) during production and consumption provide information on a product's carbon footprint. The ecological footprint argument gave rise to the idea of the carbon footprint, which is based on life cycle assessment (LCA). It is computed over one year and expressed as the quantity of carbon dioxide or other GHGs that are released in their equivalent. Global warming is the result of greenhouse gases created by human activity building up and becoming more concentrated in the atmosphere. Carbon dioxide is the primary cause of global warming and makes up almost 80% of emissions from developed nations [13]. The assessment of greenhouse gas (GHG) emissions in terms of carbon dioxide (CO_2), which can be classified as directly (primary) or indirectly (secondary) generated by an activity, is the essence of the carbon footprint. The direct assessment of CO_2 emissions from burning fossil fuels for energy and transportation purposes is known as the primary carbon footprint, or direct carbon footprint. The assessment of CO_2 emissions from the procurement of raw materials to the product disposal process is known as the secondary carbon footprint, or indirect carbon footprint (Life cycle of products and Sustainability) [14, 15]. Because they are renewable and biodegradable, natural dyes have a lower carbon impact and require fewer chemicals and energy to produce. Because synthetic dyes need more energy to manufacture and have the potential to release hazardous compounds into the environment, they have a higher carbon footprint than natural colors [16]. Fabric varieties and production methods have a significant impact on the textile industry's carbon footprint (CF). Compared to synthetic materials like polyester and nylon, which need more energy and chemicals to produce, natural fibers like cotton, wool, and silk often have a lower CF. Furthermore, compared to fabrics made conventionally, those made using eco-friendly techniques, such as closed-loop systems and sustainable production procedures, have a far lower CF [17].

The Earth's temperature has risen as a result of the greenhouse effect being intensified by rising atmospheric concentrations of greenhouse gases (GHGs). Climate change is mostly the result of human activity, particularly industrial activity. Reactive, acid, disperse, direct, naphthol, indigo, lanaset, and basic dyes are among the many types of dyes used for different kinds of fabric. The dye industry is a typical energy consumer that produces a lot of waste and emissions each year, which contribute to the global climate through greenhouse gas emissions.

Over the past 40 years, textile output has increased more than fourfold in response to the growing demand for textile products worldwide. By 2030, the number of textiles consumed worldwide is predicted to rise by as much as 148 million tons. Nevertheless, because of the textile industry's success, 0.8 million tons of synthetic dyes are consumed annually to meet the demands of textile production. Emissions of textile dyes into the atmosphere can endanger human health and contribute to air pollution. Of all the textile dyes, disperse dye is the most commonly used and is produced in China in the greatest quantity for export. Polyester, nylon, and other synthetic materials are the primary substrates for the application of disperse dyes. Analyzing the dyes' GHG emissions and determining possible ways to enhance energy-saving management in dye manufacture are therefore essential. A popular sustainable indicator for measuring the GHG impact of products and services is the life-cycle-based carbon footprint (CF). The CF is a thorough instrument for calculating GHG emissions throughout a product's whole life cycle, from the procurement of raw materials to the production and disposal of trash. It is frequently used to investigate GHG emission reduction strategies and assess the effects of various products on climate change. Necessary standardization procedures have been created to offer the specifications for measuring the GHG impact. The World Resources Institute and the World Business Council for Sustainable Development created the Product Life Cycle Accounting and Reporting Standard as one method. The British Standards Institute's PAS 2050: 2011 standard is another. Another significant technical standard for measuring, tracking, and disclosing GHG emissions is ISO 14067. The three standards differ primarily in how certain types of emissions are considered and the reporting requirements. However, these three standards are comparable in general, and they are all founded on the life cycle thinking methodology. Thus, life cycle evaluation is the methodology used in this work to evaluate the CF of the common dyes (LCA). Numerous studies have been conducted about the CF calculation in chemistry production, primarily pertaining to liquid biofuels, ethanol, aluminium, magnesium, calcium carbide, methyl crotonate, and methyl acrylate. In order to increase energy-use efficiency, earlier research qualified GHG emissions using the LCA method and identified possible energy savings. A comprehensive CF of the chemical industry's production of disperse dyes, however, has not been presented in any study. The emission factor of dyes, an essential material input in the production of textiles, ought to be considered when calculating the CF for textile products. Many national and international researchers have determined the CF for textile products. The advancement of CF theory in the textile sector, product CF accounting, and the connection between industrial CF and influencing factors have been the key topics of discussion. Silk apparel, T-shirts, cotton knits, cotton fabric, jeans, leather, yarns, and cashmere fabrics are a few examples of textile products that have been the subject of CF study. Instead of employing separate factors for each dye, the CF for the dyeing and finishing process used in organic cotton single jersey knitted fabrics uses a single dye emission factor [18].

Another noteworthy aspect of carbon footprinting is that it frequently attempts to take emissions into account across a product or service's entire life cycle. The global standard for life cycle assessment, ISO14040, serves as the foundation for both carbon footprinting and PEF calculations, which are simplified versions of each other. While a PEF or LCA frequently takes into account additional resource, environmental, and human health categories like energy usage, habitat impacts, and carcinogen emissions, carbon footprinting is simplified in that it simply takes into account one effect area (climate change) [19].

CONCLUSION

Both synthetic and natural dyes have advantages and disadvantages of their own. Natural colors are derived from renewable resources and are safer for the environment. They are perfect for traditional methods and environmentally friendly items. However, they frequently require a mordant to put the dye on fabric, can be less consistent in color, and may be more challenging to apply. Conversely, synthetic dyes are vivid, dependable, and simple to apply. They provide outstanding stability, solubility, and colorfastness. However, if not adequately controlled, their usage and manufacture frequently lead to pollution and carbon emissions, making them detrimental to the environment. This chapter promotes the use of more environmentally friendly dyeing methods by contrasting the carbon footprint, application, and reproducibility of the two types. To promote a cleaner, greener textile business, future research should concentrate on enhancing natural dyeing techniques and lessening the environmental impact of synthetic colors.

ACKNOWLEDGMENTS

The authors wish to recognize the application of AI tools, like Grammarly, during the writing process to enhance the clarity and language of the manuscript. They have thoroughly reviewed and modified the content as necessary and accept full responsibility for the content of the chapters. Additionally, they would like to emphasize that AI was not utilized in the chapters to create or modify images, aid in data analysis, or contribute to their research.

REFERENCES

[1] Saxena S, Raja A. Natural dyes: sources, chemistry, application and sustainability issues. Roadmap to sustainable textiles and clothing: eco-friendly raw materials, technologies, and processing methods. Springer 2014; pp. 37-80. [http://dx.doi.org/10.1007/978-981-287-065-0_2]

[2] Ali AN. A comprehensive study of natural and synthetic dyes: their properties, methods of preparation, and uses. SHIFAA 2024; 2024: 1-17. [http://dx.doi.org/10.70470/SHIFAA/2024/001]

[3] Bishal A, Ali KA, Ghosh S, *et al.* Natural Dyes: Its Origin, Categories and Application on Textile Fabrics in Brief. Eur Chem Bull 2023; 12(8): 9780-802.

[4] Gupta VK. Fundamentals of Natural Dyes and Its Application on textile substrates. In: Chemistry and Technology of Natural and Synthetic Dyes and Pigments. IntechOpen; 2020.

[5] Samanta AK, Awwad N, Algarni HM. Chemistry and technology of natural and synthetic dyes and pigments. IntechOpen; 2020. [http://dx.doi.org/10.5772/intechopen.83199]

[6] Ragab M, Hassabo A, Othman H. An Overview of Natural Dyes Extraction Techniques for Valuable Utilization on Textile Fabrics. Journal of Textiles, Coloration and Polymer Science 2022; 19(2): 137-53.

[7] Slama HB, Chenari Bouket A, Pourhassan Z, *et al.* Diversity of synthetic dyes from textile industries, discharge impacts and treatment methods. Appl Sci (Basel) 2021; 11(14): 6255. [http://dx.doi.org/10.3390/app11146255]

[8] Alegbe EO, Uthman TO. A review of history, properties, classification, applications and challenges of natural and synthetic dyes. Heliyon 2024; 10(13): e33646. [http://dx.doi.org/10.1016/j.heliyon.2024.e33646] [PMID: 39040261]

[9] Ali H. Biodegradation of synthetic dyes—a review. Water Air Soil Pollut 2010; 213(1-4): 251-73. [http://dx.doi.org/10.1007/s11270-010-0382-4]

[10] Harris S, Brøns C, Zuchowska M. Textiles in Ancient Mediterranean Iconography. Oxbow Books 2021.

[11] Ardila-Leal LD, Poutou-Piñales RA, Pedroza-Rodríguez AM, Quevedo-Hidalgo BE. A brief history of color, the environmental impact of synthetic dyes and removal by using laccases. Molecules 2021; 26(13): 3813. [http://dx.doi.org/10.3390/molecules26133813] [PMID: 34206669]

[12] Hassan MN, Fariha FT, Zayee Z, Hasan M. Analyzing different functional and dyeing performance of natural blended fabric utilizing natural dyes and quality prediction by fuzzy logic. Heliyon 2024; 10(22): e40399. [http://dx.doi.org/10.1016/j.heliyon.2024.e40399] [PMID: 39634399]

[13] Jain M. Ecological approach to reduce carbon footprint of textile industry. International Journal of Applied Home Science 2017; 4(7/8): 623-33.

[14] Dal A, Simsek Yesil E, Ozturk E, Kitis M. Investigation of water and carbon footprint reductions employing best available techniques in the textile sector. J Clean Prod 2024; 466: 142913. [http://dx.doi.org/10.1016/j.jclepro.2024.142913]

[15] Athalye A. Carbon footprint in textile processing. Colorage 2012; 59(12): 45-7.

[16] Mukherjee A, Sriniketan B, Bengal W. Innovative dyeing and printing with natural dyes. Book of Abstracts 2011.

[17] Leal Filho W, Dinis MAP, Liakh O, *et al.* Reducing the carbon footprint of the textile sector: an overview of impacts and solutions. Text Res J 2024; 94(15-16): 1798-814. [http://dx.doi.org/10.1177/00405175241236971]

[18] Li X, Zhu L, Ding X, Wu X, Wang L. Climate change and the textile industry: The carbon footprint of dyes. AATCC Journal of Research 2024; 11(2): 109-23.
[http://dx.doi.org/10.1177/24723444231212954]

[19] Peters GM, Svanström M, Roos S, Sandin GA, Zamani B. Carbon footprints in the textile industry. In: Muthu SS (Ed.) Handbook of life cycle assessment (LCA) of textiles and clothing. 2015, Elsevier. p. 3-30.
[http://dx.doi.org/10.1016/B978-0-08-100169-1.00001-0]

CHAPTER 4

Bio-based Textile Materials Towards Green Chemistry in the Fashion Industry

Mohammad Khajeh Mehrizi[1,*], **Shahid Adeel**[2], **Mahdieh Dehghan**[1], **Zahra Shahi**[1] and **Muhammad Usama**[3]

[1] *Department of Textile Engineering, Yazd University, Yazd, Iran*

[2] *Department of Applied Chemistry, Government College University Faisalabad, Faisalabad, Pakistan*

[3] *Department of Biochemistry, Government College University Faisalabad, Faisalabad, Pakistan*

Abstract: The fashion industry has a significant impact on the environment and society, largely due to the prevalence of fast fashion, complex supply chains, and substantial textile waste. Substituting sustainable methods, such as using eco-friendly materials and recycling, is essential to reduce pollution, resource use, and environmental damage. Textile materials made from renewable sources, such as plants and microorganisms, are a good substitute for synthetic fibers because they break down naturally and have a lower carbon footprint. Development in green chemistry, bioengineering, and new dyeing and recycling approaches also helps create more eco-friendly manufacturing procedures. Although there are challenges such as recycling mixed fibers and high expenses, continued research, industry cooperation, and effective government policies are crucial for developing a sustainable textile industry. Using these eco-friendly plans can significantly lower the industry's impact on the environment and support worldwide efforts to achieve conservation, sustainability, and responsible consumption.

Keywords: Bio-based fibers, Environmental impact, Eco-friendly, Green chemistry, Sustainable textile.

INTRODUCTION

The fashion industry is increasingly in the spotlight and plays a significant role in global environmental and social issues. Life Cycle Assessment is a standard tool used to study the ecological impacts of all stages of a product's life [1].

* **Corresponding author Mohammad Khajeh Mehrizi:** Department of Textile Engineering, Yazd University, Yazd, Iran; E-mail: mkhajeh@yazd.ac.ir

OVERVIEW OF THE FASHION INDUSTRY'S ENVIRONMENTAL IMPACT

Over the past few decades, the fashion industry has seen significant shifts in trends. It has evolved into a complex, widespread, and global network primarily focused on encouraging people to constantly buy new items and dispose of old ones. The rise of "fast fashion" encourages people to quickly replace their clothing as trends change rapidly. This business model has significant negative impacts on the environment and society, especially for workers involved in manufacturing. To reduce costs, companies often produce clothing in developing countries where labor is inexpensive and environmental and social regulations are less stringent [2, 3]. Declining clothing prices and faster production cycles, along with poor quality and planned obsolescence, have led to an increase in the global volume of consumer clothing [4]. There are significant problems with clothing waste, as most clothing and textile waste ends up in landfills.

Recycling or Reusing [2, 4]: In 2000, global clothing sales were $1 trillion, with one-third of sales in Western Europe, one-third in North America, and one-quarter in Asia [2]. This not only increases the extent of environmental impacts, but also exacerbates the issues of disposing of the large amounts of textile waste produced. In the 1990s, the fashion industry faced a public backlash due to the lack of social responsibility and accountability among factories located in developing countries [1]. Negative publicity about "sweatshop" labor pushed the fashion industry to create codes of conduct, sourcing policies, and Corporate Social Responsibility (CSR) practices [1, 5]. Like many other industries, fashion increasingly recognizes that companies have responsibilities to a wide range of stakeholders. These concepts have become integrated into the management and operation of many fashion companies, as considering stakeholders and analyzing their needs has been linked to improved organizational performance [6]. Growing awareness and stakeholder concerns have prompted the fashion industry to gradually address the environmental and social effects of its production [7]. Although "green" fashion was initially criticized in the early 1990s for being expensive and low quality, the past decade has seen the rise of "eco-fashion" (or "sustainable fashion"), which suggests a commitment to sustainability.

While terms like "ethical," "green," and "eco" lack standardized definitions within the industry, their frequent use suggests that more brands are attempting to attract mainstream customers with sustainable clothing. Successful ethical brands, such as People Tree, American Apparel, and Edun, have emerged. Major retailers, including H&M, Nike, Levi's, and Zara, have also introduced products made with eco-friendly materials such as organic cotton, Tencel, and recycled polyester [1].

Importance of Sustainability in Textiles

The clothing and textile industry is a global polluter facing complex and interconnected sustainability problems. It's a key topic in discussions about climate change, chemical use, water shortages, and human rights. The way we produce and consume textiles and clothing raises questions about our political, social, and economic lives. These challenges stem from shared social and personal habits and conflicting values related to production and consumption. Many different groups are involved, including corporations, governments, non-profit organizations, media outlets, and individuals.

While technology can solve some problems, others require action from consumers, NGOs, governments, and businesses on a global scale. Businesses and consumers are particularly important because of the nature of the textile and fast fashion industries. Policymakers and activists are considering how to encourage businesses and consumers to take responsibility, voluntarily improve sustainability, or even be required to change their behaviors. The fundamental question is how much responsibility they can take for ensuring sustainable textile and clothing production.

Researchers stress that the problems in global textile and clothing production are very complex. Beyond technological improvements, changing behaviors and habits requires changing the values associated with production and consumption, while considering diverse cultural, geographical, and political contexts [8].

Environmental sustainability is crucial for the textile and apparel industry, which makes a significant contribution to socio-economic development in developing countries [8 - 10]. For instance, cotton textile production employs seven percent of the workforce in these countries [11, 12]. However, textile production consumes substantial resources and can cause severe environmental damage. Therefore, businesses should prioritize environmental sustainability, striking a balance between economic efficiency and environmental impact [13 - 15]. The concept of environmental sustainability originates from sustainable development [16, 17], which acknowledges the interconnectedness of environmental, economic, and social sustainability [18, 19]. While environmental issues are the initial focus of environmental sustainability assessments, economic and social factors are also essential. Research emphasizes the importance of considering all these factors together when evaluating environmental sustainability [20 - 22].

The fashion industry is a major polluter, second only to the oil industry. To address this, the EU launched the "European Clothing Action Plan" in 2019. This initiative aims to make the textile industry more sustainable, from the design stage to the disposal of clothing. A significant issue is the substantial amount of textile waste that ends up in landfills worldwide. If all clothing brands adopted eco-friendly practices, the fashion industry could become much more sustainable. In fact, up to 95% of textiles currently sent to landfills each year could be recycled. (pure waste.org) As consumers become increasingly concerned about the social and environmental impacts of their purchases, businesses are starting to recognize the benefits and untapped economic potential of efficiently utilizing waste [23].

Understanding Bio-based Textile Materials

We are using more and more synthetic fibers every day. However, because these fibers rely heavily on petroleum-based chemicals, we're using up a significant amount of non-renewable resources. This dependence causes problems like:

- Petroleum shortages
- Economic and environmental issues
- Reliance on a few oil-rich countries
- Predicted depletion of oil reserves

That's why it's increasingly important to research and develop bio-based materials to reduce our use of fossil fuels. Bio-based synthetic fibers are preferable because they're made from biomass, which results in a lower carbon footprint during production. Many of these fibers also break down easily in natural environments or through microorganisms, even without special conditions. However, it's important to note that not all bio-based fibers are biodegradable. Improving biodegradability is a major goal in bio-based fiber research. In the past, bio-fibers were made from food products, but now we're shifting to non-food sources. To protect the environment, ensure sustainability, and conserve resources, it's necessary to produce non-food products and bio-based raw materials to obtain bio-based textile fibers and even develop ideal bio-based raw materials that are carbon-negative, such as those derived from moss and CO_2. These bio-based textile fibers have great potential for use in sustainable clothing and medical textiles due to their non-toxic, skin-friendly, and antibacterial properties [24].

Types of Bio-based Fibers

Burning synthetic fibers, a common disposal method, releases significant greenhouse gases into the atmosphere [25]. In contrast, polylactic acid (PLA) fiber breaks down naturally in soil through a process called hydrolysis. Microorganisms, such as Actinomycetes, further accelerate this biodegradation [26]. Plants absorb the carbon dioxide released during PLA's decomposition, becoming part of the soil and preventing its release into the air, thereby avoiding any contribution to the greenhouse effect. This makes PLA fiber an environmentally friendly alternative. Consequently, PLA is considered a leading biodegradable material and is widely used in various industries, including food packaging, transportation, textiles, and electronics. Polyhydroxyalkanoates (PHAs) are another type of biodegradable material. These are natural polyesters produced within microorganisms. PHAs can be broken down by microorganisms in various environments, including compost, soil, and seawater. The resulting by-products are primarily water and carbon-based, causing no pollution [27 - 29].

Early bio-based polymers for textiles came from agricultural sources. These "first-generation" materials included starch from crops like corn, wheat, potatoes, sugar beets, and rice, as well as vegetable oils. Sugar, often from sucrose, was fermented and converted into polymers [30, 31]. For example, corn starch can be processed into glucose, which is then fermented using specific strains to produce pure lactic acid. This lactic acid is chemically synthesized into polylactic acid (PLA), a polymer known for its biodegradability, bioabsorption, and biocompatibility. Producing PLA requires 75.4 MJ/kg of energy. This relatively low energy consumption, along with low gas emissions, makes PLA production more efficient. Compared to traditional petroleum-based polymers, the production of PLA reduces fossil energy consumption by 25-55% [32]. Earhart *et al.* (2012) explored the use of fructose from corn to create bioplastic polyethylene furan dicarboxylate (PEF). Their research on energy and greenhouse gas emissions showed that PEF production could lower non-renewable energy consumption by 40-50% and greenhouse gas emissions by 45-55% compared to producing PET (polyethylene terephthalate) [33].

The first generation of bio-fibers, while useful, presents challenges. Producing them from food crops can lead to food shortages and disrupt the balance between food supply and population growth due to economic pressures. Furthermore, current trends suggest that the availability of starch and sugar, which are essential for these biofibers, will be limited in the future [24].

The focus is shifting towards using non-food parts of plants as bio-based raw materials [34]. This "second generation" of bio-based materials involves processing carbohydrate polymers, such as cellulose and lignin, extracted from non-food sources. Examples include bagasse, corn waste, palm oil waste, and other agricultural waste [35].

Using these non-food sources avoids impacting the food supply while utilizing waste materials. For example, after corn is harvested for food, the remaining stalks are often used as animal feed or burned. However, corn stalks can also be used to produce chemicals like furfural, xylose, and xylitol [36]. Corn stalks are particularly suitable for this because of their high xylan content and low molecular crystallinity, which makes them easily broken down by enzymes [37]. Lignin, the second most abundant natural polymer in plants, is a major renewable source of aromatic compounds [38]. It is considered an ideal raw material for producing bio-based aromatic chemicals. Currently, the paper industry generates approximately 50 million tons of lignin annually as a by-product; however, 98% of it is burned [39, 40]. This underutilizes lignin's potential due to its abundant aromatic rings. Researchers are exploring ways to utilize lignin more effectively.

Third-generation biopolymers, derived from diverse natural sources such as moss, CO2, and bacterial cellulose, hold significant potential; however, their development and application are currently limited by technological constraints. Sodium chlorite pretreatment significantly enhances enzymatic hydrolysis, increasing glucose yield to 70% compared to 5% without pretreatment, indicating a high potential for bioethanol production.

Sources of Bio-based Materials

Microorganisms have been used to produce food products, such as vinegar, yogurt, and cheese, for over 8,000 years. A century ago, French merchants sought to prevent their wine and beer from souring during long shipments and turned to Louis Pasteur for assistance. Pasteur discovered that yeasts convert sugar into alcohol through anaerobic fermentation. He also identified that the sourness and spoilage result from acetic acid bacteria transforming alcohol into vinegar.

The term biotechnology was first used in 1919 by Karl Ereky to mean the application of medical and biological knowledge and its opposite effect in man-made technologies. In general, any intelligent human action in the creation, improvement, and supply of various products using living organisms, especially through their manipulation at the molecular level, falls within the scope of this most important, clean, and economical technology of the current century: biotechnology. The name of this knowledge is therefore known as "biotechnology" in Iran, and the same name is known for this practice in almost all countries of the world.

The use of biomass to produce commodity chemicals continues to grow. End uses such as the automotive, housing, and consumer goods industries, including textiles, as well as infrastructure, produce chemical intermediates and commodity chemicals from renewable resources. The production of basic chemicals from biomass is not a new concept, and interest in it has waned over the past century. With changing international priorities and advances in chemistry and biology, the potential to utilize biomass or other renewable resources, such as natural oils and even carbon dioxide, is more promising than ever [41].

GREEN CHEMISTRY PRINCIPLES IN TEXTILE PRODUCTION

Definition and Importance

Green Chemistry is defined as the thoughtful and systematic design of chemical products and processes with the primary aim of reducing or completely eliminating the presence of hazardous substances that may pose risks to health and the environment. This approach emphasizes the importance of creating safer chemicals and processes that minimize harmful effects throughout their life cycle. This concept emerged in the early 1990s, nearly 20 years ago [42]. Since then, it has gained international traction, leading to the establishment of hundreds of programs and governmental initiatives worldwide, particularly in the United States, the United Kingdom, and Italy, significantly advancing sustainable design [43].

The cornerstone of green chemistry is intentional design, characterized by novelty, planning, and systematic thinking. The twelve principles of green chemistry act as "design rules" to guide chemists towards sustainability. It prioritizes careful planning in chemical synthesis and molecular design to reduce negative impacts.

The green chemistry approach promotes molecular-level sustainability, leading to its widespread use across diverse industries. Successful and economically competitive technologies can be found in various industries, including aerospace, automotive, cosmetics, electronics, energy, home goods, pharmaceuticals, and agriculture [44].

Green chemistry has significantly influenced various sectors, including research laboratories, industrial operations, educational institutions, environmental practices, and public health initiatives. It clearly illustrates how chemists can creatively design and produce groundbreaking products and processes that not only drive profitability but also contribute positively to human health and protect the environment. By prioritizing sustainability and reducing harmful substances, green chemistry empowers scientists to devise solutions that are both economically viable and ecologically sound, demonstrating its crucial role in fostering innovation for the betterment of society as a whole.

The key topics of the green chemistry outline are:

1. It integrates design throughout the entire chemical life cycle.
2. It aims to mitigate the inherent hazards associated with chemical products and processes.

3. It operates as a unified system of design principles.

Green chemistry aims to minimize hazards throughout all life-cycle stages while demonstrating economic benefits. Hazards refer to the potential harm to humans or the environment. The intrinsic risks associated with chemicals—such as toxicity, physical dangers like explosions and flammability, and global issues like ozone depletion—can be mitigated at every stage of the process. In chemical transformations, these risks might originate from the feedstock, raw materials, and end products. By integrating the twelve principles, thoughtful design can successfully lessen or eliminate these risks.

The basic principles of green chemistry encompass a range of synthetic organic production issues, with an emphasis on developing procedures that minimize waste and by-products, reduce the use of hazardous chemicals, and promote the use of safer solvents. Green chemistry ensures that we mitigate adverse effects on ecosystems and promote a more sustainable future by emphasizing these approaches [45].

Principles of Green Chemistry

In 1998, Paul Anastas and John Warner identified the twelve principles of green chemistry, which provide a framework for developing safer chemical processes and products. These principles aim to minimize environmental and human health hazards by reducing or eliminating toxic materials throughout the synthesis, production, and chemical usage processes. The goal of green chemistry is to incorporate as many of the twelve principles as feasible into the synthesis process, even though doing so can be challenging [45].

Prevention

Preventing waste is preferable to managing it after it has been created. The first principle of green chemistry emphasizes waste prevention, which benefits both humans and the environment. It is ultimately more cost-effective than treating or disposing of waste after it is generated.

Atom Economy

Synthetic methods should aim to maximize the incorporation of material into the final product. Atom economy, which focuses on the complete utilization of raw materials, is closely tied to waste prevention. Therefore, chemical synthesis should be designed to ensure that the final product fully utilizes all inputs.

Less Hazardous Chemical

In the development of synthetic procedures, it is essential that, wherever feasible, the methods employed are carefully designed to utilize substances that not only have minimal toxicity but also generate products that are similarly non-toxic to both human health and the surrounding environment. This approach aims to ensure that both the materials used in synthesis and those produced as a result are safe, thereby reducing the potential risks associated with hazardous chemicals during the synthesis process.

Designing Safer Chemicals

Designing Safer Chemicals promotes the creation of chemical products that minimize toxicity while maintaining effectiveness. The objective is to achieve a balance between optimal performance and safe usage, ensuring that risks associated with toxic substances (non-carcinogenic, mutagenic, neurotoxic) are minimized. Toxic chemicals should be avoided and substituted whenever feasible, without compromising their efficacy.

Safer Solvents and Auxiliaries

Safer solvents and auxiliaries recommend minimizing synthesis processes and avoiding auxiliary chemicals, such as solvents and separation agents, whenever feasible. If their use is necessary, they should be non-harmful. Following green chemistry principles, alternatives to organic solvents should prioritize worker and process safety, environmental protection, and sustainability. The ideal solvent should be stable, low in volatility, user-friendly, and easily recyclable.

Design for Energy Efficiency

The energy supplies associated with various chemical procedures need to be carefully acknowledged due to their significant environmental and economic implications. It is crucial to recognize that these processes can contribute to a range of adverse effects, both environmentally and economically. Therefore, efforts should be made to minimize these energy demands whenever possible. In addition, if feasible, synthetic methods should ideally be conducted at ambient temperature and pressure. This approach not only helps reduce energy consumption but also aligns better with sustainable practices that are less harmful to the environment. By prioritizing such conditions, we can further enhance the efficiency and sustainability of chemical processes while limiting their overall impact.

Use of Renewable Feedstocks

Using renewable raw materials is more convenient than relying on various plastics and discarding waste. Consequently, the manufacture of biodegradable plastics is a growing trend, particularly in the food industry. Several factors, including political changes, legislation, and the global demand for food and energy resources, influence the development of biodegradable packaging.

Reduce Derivatives

A key principle of green chemistry in synthesizing target molecules is to minimize the use of chemical derivatives. This encourages avoiding processes that require blocking and unblocking groups, and wherever possible, favors biological synthesis to prevent the production of compounds that lack specific degrading enzymes.

Catalysis

Catalytic reagents are preferred over stoichiometric ones due to their selectivity. To benefit the environment, catalysis promotes the use of biodegradable catalysts that consume less energy, avoid organochlorine compounds, and reduce water usage and waste.

Design for Degradation

Chemical products should be designed to degrade into harmless substances after use, preventing environmental persistence. This principle emphasizes creating chemicals that safely transform after fulfilling their purpose.

Real-Time Analysis for Pollution Prevention

Real-time analysis for pollution prevention requires the development of continuous monitoring methods for chemical production. This strategy aims to prevent hazardous substance formation by ensuring constant oversight of each production stage, thereby minimizing errors that could pose environmental and health risks.

Inherently Safer Chemistry for Accident Prevention

The twelfth principle of green chemistry is fundamentally safer chemistry for accident prevention, which emphasizes minimizing the use of hazardous substances in chemical processes that can lead to explosions, fires, or the release of harmful vapors. A current instance is the growing use of supercritical CO_2 as a replacement for organic solvents, as it is non-toxic, non-explosive, and environmentally friendly.

Disadvantages of Green Chemistry

The primary objective of green chemistry is to design chemical products and processes that minimize or eliminate the use of harmful substances. However, this goal presents significant challenges, including time constraints, high costs, and insufficient information. Transitioning from traditional products or processes to greener alternatives often requires extensive time and effort, as redesigning these systems can be complex and costly. Additionally, there is a lack of consensus on what constitutes safety in this context. The high implementation costs and insufficient information hinder the adoption of green chemistry, as viable alternatives to existing chemical raw materials and technologies for green procedures remain unknown. Additionally, there is a shortage of skilled human resources, and the risks associated with transitioning to green products and processes are not evenly distributed throughout the supply chain. Furthermore, there are limited resources for further research.

Future of Green Chemistry

Many chemical processes today need to become more environmentally friendly, following the principles of green chemistry. Current methods for making many products don't fully align with these principles. They often use crude oil, which creates a significant amount of waste. Adopting green chemistry principles can be challenging, but focusing on them can lead to important new discoveries and advancements in chemistry. This shift in how we practice chemistry is not only environmentally necessary but also an opportunity for scientists to reassess how we create and utilize materials. For example, the use of biocatalysis and alternative raw materials is becoming increasingly popular. These approaches help reduce our confidence in fossil fuels and minimize hazardous waste. Through hopeful collaboration between experts in chemical engineering, biology, and materials science, researchers can discover new ways to promote sustainability.

Furthermore, education plays a crucial role in shaping future chemists who prioritize the principles of green chemistry. By integrating these considerations into curricula, universities and research centers can cultivate a generation of scientists equipped with the knowledge and tools to make meaningful changes. Engaging students in sustainable practices not only increases their awareness of environmental impacts but also empowers them to tackle complex problems with innovative solutions. Together, industry partnerships and government policies can create incentives for adopting greener practices in manufacturing and product development. By making clear regulations and providing support for research funding, policymakers can accelerate the transition to more sustainable chemical processes.

Through continued research, education, and collaboration, the chemical industry can evolve in a way that not only respects the planet but also enhances human health and well-being. Ultimately, striking a balance between innovation and environmental stewardship will be key to overcoming the challenges ahead.

INNOVATIONS IN BIO-BASED TEXTILE DEVELOPMENT

The fundamental development in bio-based textiles, such as bio-dyes, has promoted the application of biomaterials in the textile industry, which reduces the negative impact of the clothing sector on our ecosystem. Nowadays, researchers are industrializing bio-fabricated products to contribute to a safer and cleaner ecosystem. Scientists use bacterial cellulose to produce textile products by cultivating bacteria through fermentation. This method allows bacteria to grow and produce bio-fabricated material with the same properties as leather. Previously, scientists successfully redefined fruit waste to generate bio-based materials. They extract the fiber and pectin from fruit leftovers and manufacture leather-like sheets using a bio-binder [46].

The most notable progress in the development of bio-based materials is the creation of recyclable textile goods synthesized from a mixture of organic cotton, algae-based compounds, Pinatex, and Tencel, which could be a promising alternative to synthetic fibers. The textile industry is now shifting back to natural dyeing approaches. CO2 dyeing and digital printing are modern techniques that help protect the environment from carcinogenic effluents and greenhouse gases [47]. The development of innovative and advanced methodologies aims to achieve zero-waste production, thereby improving effectiveness in resource management.

In recent years, the production of homo- and co-polymers using a new type of furan ring-based biopolymer synthesized from 2,5-furandicarboxylic acid (FDCA) has garnered notable attention from manufacturers and researchers [48]. These biopolymers are not widely used due to economic constraints [49]. Due to their excellent thermal and mechanical properties, poly(alkylene furanoate) (PAF) is a widely studied biopolymer. PEF (polyethylene 2,5 furan-dicarboxylate) is considered the most beneficial biodegradable PAF, which is comparable to polyethylene terephthalate (PET) [50]. Compared to PET, PEF exhibits a more enhanced gas barrier potential, as well as excellent mechanical and thermal resistance, and could be a promising alternative to PET [51, 52]. Hence, combining these innovative biopolymers with existing bio-based plastic materials could offer remarkable properties. In a polymer blend, two polymers are combined to produce a new bio-based material that exhibits different yet improved physical characteristics [53].

ENVIRONMENTAL IMPACT AND CONCERNS ASSOCIATED WITH BIO-BASED FIBERS

The textile industry has contaminated our ecosystem by emitting greenhouse gases, polluting effluent, and releasing hazardous chemicals. To fulfill the preferences and tastes of consumers, the textile industry boosts the production of clothing and accessories. These textile goods have resulted in textile waste that is often burned or disposed of in landfills, posing significant environmental problems [54].

Hence, replacing man-made materials with bio-based alternatives could make the textile sector more sustainable and greener. The advancements and innovations in recent technologies have been made to promote sustainable, organic, and greener products, as well as to alleviate the negative impact of the textile sector on aquatic life and the environment. Moreover, such sustainable innovations in the apparel industry make the textile sector eco-friendlier and greener [55]. Therefore, these advancements influence the production of textile goods, their quality, consumer experience, and the supply of materials aimed at reducing waste production.

Recently, there has been an increase in research and investigations aimed at developing recyclable textile products using organic and biodegradable raw materials. For example, conventional cotton production, combined with synthetic polyester, consumes a large quantity of water and energy, making the dyeing process unsustainable. Hence, innovations in dyeing methodologies, such as the CO_2 dyeing approach, use half the energy and no water compared to conventional dyeing approaches [56]. On the other hand, the revival of natural dyes has replaced synthetic colorants, making the dyeing process eco-friendly and alleviating the negative impact of the textile sector on our ecosystem.

The environmental impact assessment of bio-composite synthesized from chicken feathers and polylactide was investigated in a previous study [57]. Cradle-to-gate life-cycle inventories were investigated, and the outcomes revealed that combining bio-fiber with a polylactide matrix reduced the environmental impact. In another study, researchers investigated the cost and environmental assessment of polypropylene nano-composites. The results showed that the nano-composite combined with plant fiber has notable environmental advantages throughout the entire life cycle [58].

The breakdown of cellulose in plant fiber not only depends on bond breaking but also on the degree of polymerization, which makes the degradation process more difficult [59]. In a previous study, the environmental impact was investigated to enhance the production process of a bio-composite made up of plant-based bio-fiber [poly (b-hydroxybutyrate)-co-(b-hydroxyvalerate)] [60]. Results showed that the addition of plant-based fiber reduces both energy consumption and the emission of greenhouse gases.

The bio-based fiber, manufactured from renewable natural resources such as animals and plants, has key advantages over synthetic fibers. These bio-based fibers originate from natural resources, so they decompose naturally and do not pollute our environment. The synthesis of these fibers from natural resources reduced the consumption of fossil fuels. During the production of bio-based fiber, the emission of greenhouse gases was recorded as very low compared to the emission during the production of synthetic fibers, such as nylon [61]. Moreover, bio-based fibers require minimal processing and consume less water and energy resources as compared to man-made fibers. These bio-based fibers are biodegradable and do not produce hazardous residues after degradation.

Bio-based fibers are more recyclable in their pure form than synthetic fibers. Natural fibers, such as wool, cotton, silk, jute, hemp, and flax, are recycled by transforming them into fibers and then spun back into new yarn or used in non-woven products. The recycling of natural fibers causes a reduction in their tensile strength and durability. However, chemical recycling of natural fibers is under investigation, in which natural fibers are converted into their base molecules and then regenerated, which could be a more beneficial solution [56]. The recycling of bio-fibers becomes more limited when they are blended with synthetic fibers, dyed with non-biodegradable colorants, or treated with a high quantity of reactive chemicals and additives.

CONCLUSION

The recent advancement in the textile sector may help provide cost-effective production, limit waste production, and reduce the emission of greenhouse gases into the environment. The modern textile industry has introduced the concept of repairing, reselling, and renting models to reduce the growing load on the textile sector and promote the recycling of textile waste. Bio-based fibers are promising alternatives to synthetic fibers due to their recyclability and biodegradability, which reduce the adverse impact of the textile sector on our ecosystem. Bio-based fibers can play a key role in environmental sustainability by implementing eco-friendly and sustainable farming approaches for the production of bio-based raw material. The advancement in sustainable processing technology and methodology also aimed to enhance the recyclability of textile products.

ACKNOWLEDGMENTS

The authors wish to recognize the application of AI tools, like Grammarly, during the writing process to enhance the clarity and language of the manuscript. The authors thoroughly reviewed and modified the content as necessary and accept full responsibility for the content of the chapters. Additionally, they want to emphasize that AI was not utilized in their chapters to create or modify images, aid in data analysis, or contribute to their research.

REFERENCES

[1] Kozlowski A, Bardecki MJ, Searcy C. Environmental Impacts in the Fashion Industry. The Journal of Corporate Citizenship 2012: 16-36.

[2] Allwood JM, Laursen SE, De Rodríguez CM, Bocken NM. Well dressed. The present and future sustainability of clothing and textiles in the United Kingdom. 2006.

[3] Hethorn J, Ulasewicz C. Sustainable fashion: why now?: a conversation about issues, practices, and possibilities. Fashion Practice 2008; 1(1): 135-6.

[4] Cohen-Cole E, Fletcher JM. Is obesity contagious? Social networks *vs*. environmental factors in the obesity epidemic. J Health Econ 2008; 27(5): 1382-7. [http://dx.doi.org/10.1016/j.jhealeco.2008.04.005] [PMID: 18571258]

[5] Park H, Lennon SJ. The organizational factors influencing socially responsible apparel buying/sourcing. Cloth Text Res J 2006; 24(3): 229-47. [http://dx.doi.org/10.1177/0887302X06293070]

[6] Donaldson T, Preston LE. The stakeholder theory of the corporation: Concepts, evidence, and implications. Acad Manage Rev 1995; 20(1): 65-91. [http://dx.doi.org/10.2307/258887]

[7] Chen HL, Burns LD. Environmental analysis of textile products. Cloth Text Res J 2006; 24(3): 248-61. [http://dx.doi.org/10.1177/0887302X06293065]

[8] Boström M, Micheletti M. Introducing the sustainability challenge of textiles and clothing. J Consum Policy (Dordr) 2016; 39(4): 367-75. [http://dx.doi.org/10.1007/s10603-016-9336-6]

[9] Zhao H, Lin B. Resources allocation and more efficient use of energy in China's textile industry. Energy 2019; 185: 111-20. [http://dx.doi.org/10.1016/j.energy.2019.06.173]

[10] Liu X, Klemeš JJ, Varbanov PS, Čuček L, Qian Y. Virtual carbon and water flows embodied in international trade: a review on consumption-based analysis. J Clean Prod 2017; 146: 20-8. [http://dx.doi.org/10.1016/j.jclepro.2016.03.129]

[11] Ellen MacArthur Foundation, (2017). A New Textiles Economy: Redesigning Fashion's Future. Available at: http://www.ellenmacarthurfoundation.org/publications

[12] World Wildlife Fund (WWF), (2013). Cleaner. Impacts and Better Management Practices, Greener Cotton. Available at: https://www.worldwildlife.org/industries/cotton

[13] Jia F, Yin S, Chen L, Chen X. The circular economy in the textile and apparel industry: A systematic literature review. J Clean Prod 2020; 259: 120728. [http://dx.doi.org/10.1016/j.jclepro.2020.120728]

[14] Janaina AK, Miguel P Jr, Davi BG, Barrella W. Textile sustainability: A Brazilian etiquette issue. Environ Sci Policy 2020; 109: 125-30.

[http://dx.doi.org/10.1016/j.envsci.2020.02.025]

[15] Thorisdottir TS, Johannsdottir L. Sustainability within fashion business models: A systematic literature review. Sustainability (Basel) 2019; 11(8): 2233.
[http://dx.doi.org/10.3390/su11082233]

[16] Kates RW, Clark WC, Corell R, *et al.* Sustainability Science. Science 2001; 292(5517): 641-2.
[http://dx.doi.org/10.1126/science.1059386] [PMID: 11330321]

[17] Organisation for Economic Co-operation and Development (OECD), 2004. Measuring Sustainable Development: Integrated Economic, Environmental and Social Aspects. France, Paris.

[18] Cheng R, Li W, Lu Z, Zhou S, Meng C. Integrating the three-line environmental governance and environmental sustainability evaluation of urban industry in China. J Clean Prod 2020; 264: 121554.
[http://dx.doi.org/10.1016/j.jclepro.2020.121554]

[19] Moldan B, Janoušková S, Hák T. How to understand and measure environmental sustainability: Indicators and targets. Ecol Indic 2012; 17: 4-13.
[http://dx.doi.org/10.1016/j.ecolind.2011.04.033]

[20] Liu Y, Zhu L, Zhang C, Ren F, Huang H, Liu Z. Life cycle assessment of melange yarns from the manufacturer perspective. Int J Life Cycle Assess 2020; 25(3): 588-99.
[http://dx.doi.org/10.1007/s11367-019-01705-8]

[21] Wang H, Pan C, Wang Q, Zhou P. Assessing sustainability performance of global supply chains: An input-output modeling approach. Eur J Oper Res 2020; 285(1): 393-404.
[http://dx.doi.org/10.1016/j.ejor.2020.01.057]

[22] Thies C, Kieckhäfer K, Spengler TS, Sodhi MS. Operations research for sustainability assessment of products: A review. Eur J Oper Res 2019; 274(1): 1-21.
[http://dx.doi.org/10.1016/j.ejor.2018.04.039]

[23] Moorhouse D, Moorhouse D. Sustainable design: Circular economy in fashion and textiles. Design Journal 2017; 20(sup1): S1948-59.
[http://dx.doi.org/10.1080/14606925.2017.1352713]

[24] Tian W, Huang K, Zhu C, *et al.* Recent progress in biobased synthetic textile fibers. Front Mater 2022; 9: 1098590.
[http://dx.doi.org/10.3389/fmats.2022.1098590]

[25] Oberoi IS, Rajkumar P, Das S. Disposal and recycling of plastics. Mater Today Proc 2021; 46: 7875-80.
[http://dx.doi.org/10.1016/j.matpr.2021.02.562]

[26] Qi X, Ren Y, Wang X. New advances in the biodegradation of Poly(lactic) acid. Int Biodeterior Biodegradation 2017; 117: 215-23.
[http://dx.doi.org/10.1016/j.ibiod.2017.01.010]

[27] Jendrossek D, Schirmer A, Schlegel HG. Biodegradation of polyhydroxyalkanoic acids. Appl Microbiol Biotechnol 1996; 46(5-6): 451-63.
[http://dx.doi.org/10.1007/s002530050844] [PMID: 9008883]

[28] Dilkes-Hoffman LS, Lant PA, Laycock B, Pratt S. The rate of biodegradation of PHA bioplastics in the marine environment: A meta-study. Mar Pollut Bull 2019; 142: 15-24.
[http://dx.doi.org/10.1016/j.marpolbul.2019.03.020] [PMID: 31232288]

[29] Meereboer KW, Misra M, Mohanty AK. Review of recent advances in the biodegradability of polyhydroxyalkanoate (PHA) bioplastics and their composites. Green Chem 2020; 22(17): 5519-58.
[http://dx.doi.org/10.1039/D0GC01647K]

[30] Babu RP, O'Connor K, Seeram R. Current progress on bio-based polymers and their future trends. Prog Biomater 2013; 2(1): 8-16.
[http://dx.doi.org/10.1186/2194-0517-2-8] [PMID: 29470779]

[31] Harmsen PFH, Hackmann MM, Bos HL. Green building blocks for bio-based plastics. Biofuels Bioprod Biorefin 2014; 8(3): 306-24. [http://dx.doi.org/10.1002/bbb.1468]

[32] Rajeshkumar G, Arvindh Seshadri S, Devnani GL, *et al.* Environment friendly, renewable and sustainable poly lactic acid (PLA) based natural fiber reinforced composites – A comprehensive review. J Clean Prod 2021; 310: 127483. [http://dx.doi.org/10.1016/j.jclepro.2021.127483]

[33] Eerhart AJJE, Faaij APC, Patel MK. Replacing fossil based PET with biobased PEF; process analysis, energy and GHG balance. Energy Environ Sci 2012; 5(4): 6407-22. [http://dx.doi.org/10.1039/c2ee02480b]

[34] Hajime N, Peter D, Katja L. The recent developments in biobased polymers toward general and engineering applications: Polymers that are upgraded from biodegradable polymers, analogous to petroleum-derived polymers, and newly developed. Polymers 2017; 9(10): 523.

[35] Deutschmann R, Dekker RFH. From plant biomass to bio-based chemicals: Latest developments in xylan research. Biotechnol Adv 2012; 30(6): 1627-40. [http://dx.doi.org/10.1016/j.biotechadv.2012.07.001] [PMID: 22776161]

[36] Wang A, Wang Y, Jiang T, Li L, Ma C, Xu P. Production of 2,3-butanediol from corncob molasses, a waste by-product in xylitol production. Appl Microbiol Biotechnol 2010; 87(3): 965-70. [http://dx.doi.org/10.1007/s00253-010-2557-8] [PMID: 20376634]

[37] Delgado-Arcaño Y, Mandelli D, Carvalho WA, Pontes LAM. Valorization of corncob by hydrolysis-hydrogenation to obtain xylitol under mild conditions. Waste Biomass Valoriz 2021; 12(9): 5109-20. [http://dx.doi.org/10.1007/s12649-021-01348-7]

[38] Raj T, Chandrasekhar K, Naresh Kumar A, Kim SH. Lignocellulosic biomass as renewable feedstock for biodegradable and recyclable plastics production: A sustainable approach. Renew Sustain Energy Rev 2022; 158: 112130. [http://dx.doi.org/10.1016/j.rser.2022.112130]

[39] Mahmood N, Yuan Z, Schmidt J, Xu CC. Depolymerization of lignins and their applications for the preparation of polyols and rigid polyurethane foams: A review. Renew Sustain Energy Rev 2016; 60: 317-29. [http://dx.doi.org/10.1016/j.rser.2016.01.037]

[40] Souza R, de Souza RE, Brito EO, Costa Lelis RC. A review on lignin sources and uses. J Appl Biotechnol Bioeng 2020; 7(3): 100-5.

[41] Petersen GR, Fitzgerald ND. The Changing Landscape: A History and Evolution of Bio-based Products. In: Snyder SW (Ed.) Commercializing Biobased Products: Opportunities, Challenges, Benefits, and Risks, The Royal Society of Chemistry, 2015, ch. 2, pp. 8-24. [http://dx.doi.org/10.1039/9781782622444-00008]

[42] Anastas PT, Williamson TC. In green chemistry: designing chemistry for the environment. Washington, DC: American Chemical Series Books 1996; pp. 1-20. [http://dx.doi.org/10.1021/bk-1996-0626]

[43] McDonough W, Braungart M, Anastas PT, Zimmerman JB. Applying the principles of Green Engineering to cradle-to-cradle design. Environ Sci Technol 2003; 37(23): 434A-41A. [http://dx.doi.org/10.1021/es0326322] [PMID: 14700308]

[44] Anastas P, Eghbali N. Office of Pollution Prevention and Toxics, The Presidential Green Chemistry Challenge, Award Recipients, (1996–2009). US Environmental Protection Agency, Washington, DC, EPA 744K09002.

[45] Ivanković A, Dronjić A, Martinović Bevanda A, Talić S. Review of 12 Principles of Green Chemistry in Practice. Int J Sustain Green Energy 2017; 6(3): 39-48. [http://dx.doi.org/10.11648/j.ijrse.20170603.12]

[46] Seidu RK, Eghan B, Acquaye R. A review of circular fashion and bio-based materials in the fashion industry. Circular Economy and Sustainability 2024; 4(1): 693-715. [http://dx.doi.org/10.1007/s43615-023-00303-z]

[47] Rognoli V, Petreca B, Pollini B, Saito C. Materials biography as a tool for designers' exploration of bio-based and bio-fabricated materials for the sustainable fashion industry. Sustainability 2022; 18(1): 749-72. [http://dx.doi.org/10.1080/15487733.2022.2124740]

[48] Sousa AF, Vilela C, Fonseca AC, *et al.* Biobased polyesters and other polymers from 2,5-furandicarboxylic acid: a tribute to furan excellency. Polym Chem 2015; 6(33): 5961-83. [http://dx.doi.org/10.1039/C5PY00686D]

[49] Papageorgiou GZ, Papageorgiou DG, Terzopoulou Z, Bikiaris DN. Production of bio-based 2,5-furan dicarboxylate polyesters: Recent progress and critical aspects in their synthesis and thermal properties. Eur Polym J 2016; 83: 202-29. [http://dx.doi.org/10.1016/j.eurpolymj.2016.08.004]

[50] Terzopoulou Z, Papadopoulos L, Zamboulis A, Papageorgiou DG, Papageorgiou GZ, Bikiaris DN. Tuning the properties of furandicarboxylic acid-based polyesters with copolymerization: A review. Polymers (Basel) 2020; 12(6): 1209. [http://dx.doi.org/10.3390/polym12061209] [PMID: 32466455]

[51] Sajid M, Zhao X, Liu D. Production of 2,5-furandicarboxylic acid (FDCA) from 5-hydroxymethylfurfural (HMF): recent progress focusing on the chemical-catalytic routes. Green Chem 2018; 20(24): 5427-53. [http://dx.doi.org/10.1039/C8GC02680G]

[52] Miah Mohammad Raza, Ding Jiheng, Zhao Hongran, *et al.* Enhancing the mechanical and barrier properties of biobased polyester incorporated with carboxylated cellulose nanofibers. Materials Today Communication 2024; 38: 108538. [http://dx.doi.org/10.1016/j.mtcomm.2024.108538]

[53] Gandini A, Silvestre AJD, Neto CP, Sousa AF, Gomes M. The furan counterpart of poly(ethylene terephthalate): An alternative material based on renewable resources. J Polym Sci 2009; 47(3): 295-8.

[54] Ramesh M, Deepa C, Kumar LR, Sanjay MR, Siengchin S. Life-cycle and environmental impact assessments on processing of plant fibres and its bio-composites: A critical review. J Ind Text 2022; 51(4_suppl) (Suppl.): 5518S-42S. [http://dx.doi.org/10.1177/1528083720924730]

[55] Weiss M, Haufe J, Carus M, *et al.* A review of the environmental impacts of biobased materials. J Ind Ecol 2012; 16(s1): S169-81. [http://dx.doi.org/10.1111/j.1530-9290.2012.00468.x]

[56] Rehman NU, Ullah KS, Sajid M, Ihsanullah I, Waheed A. Preparation of sustainable composite materials from bio-based domestic and industrial waste: Progress, problems, and prospects—a review. Adv Sustain Syst 2024; 8(8): 2300587. [http://dx.doi.org/10.1002/adsu.202300587]

[57] Molins G, Alvarez MD, Garrido N, *et al.* Environmental impact assessment of polylactide (PLA)/chicken feathers biocomposite materials. J Polym Environ 2017; 26: 1-12.

[58] Roes AL, Marsili E, Nieuwlaar E, Patel MK. Environmental and cost assessment of a polypropylene nano-composite. J Polym Environ 2007; 15(3): 212-26. [http://dx.doi.org/10.1007/s10924-007-0064-5]

[59] He JH, Ain QT. New promises and future challenges of fractal calculus: From two-scale thermodynamics to fractal variational principle. Therm Sci 2020; 24(2 Part A): 659-81. [http://dx.doi.org/10.2298/TSCI200127065H]

[60] Miller SA, Billington SL, Lepech MD. Improvement in environmental performance of poly(β-

hydroxybutyrate)-co-(β-hydroxyvalerate) composites through process modifications. J Clean Prod 2013; 40: 190-8.
[http://dx.doi.org/10.1016/j.jclepro.2012.08.033]

[61] Liu Fuyao, Pan Liang, Liu Yifan, *et al.* Biobased fibers from natural to synthetic: Processing, manufacturing, and application. Matter 2024; 7(6): 1977-2010.

CHAPTER 5

Valorization of Tea Waste for Sustainable Fabric Dye

Belal Uddin Rabbi[1] and **Sultana Bedoura**[2,*]

[1] *Department of Wet Process Engineering, Bangladesh University of Textiles (BUTEX), Dhaka, Bangladesh*

[2] *Department of Dyes and Chemical Engineering, Bangladesh University of Textiles (BUTEX), Dhaka, Bangladesh*

Abstract: The textile sector is significantly dependent on synthetic dyes, which lead to considerable negative environmental impacts, including water contamination, the accumulation of hazardous chemicals, and health risks for workers. With the increasing demand for environmentally friendly and sustainable dyeing methods, natural dyes from waste materials present promising alternatives. This chapter presents a comprehensive study of transforming tea waste of black, oolong, white, and green leaves into a sustainable fabric dye, serving as a viable alternative to synthetic dyes within the textile industry. It reviews the chemical constituents of tea leaves, compares different extraction methods, and analyzes key process parameters that influence color yield. The chapter evaluates key extraction parameters that influence dye yield and color intensity, providing guidance on how to optimize these variables for improved performance. The application of extracted dye on textile substrates, along with its limitations and future potential, is also discussed. Additionally, recent advances such as far-infrared and gamma irradiation pretreatments are highlighted for their potential to enhance pigment availability and improve extract color characteristics, offering new opportunities for strengthening tea-waste-based dye systems. Overall, the findings indicate that optimized tea-waste dyes provide stable coloration and offer a viable pathway toward sustainable textile production.

Keywords: Aqueous extraction, Antioxidant, Eco-friendly dyeing, Dyeing using spent tea, Phenolic compounds, Tea.

INTRODUCTION

The textile industry has recently transitioned towards sustainable dyeing practices in response to the release of highly toxic chemicals, synthetic dyes, and hazardous metallic elements into water from traditional dyeing methods. Each year, the con-

* **Corresponding author Sultana Bedoura:** Department of Dyes and Chemical Engineering, Bangladesh University of Textiles (BUTEX), Dhaka, Bangladesh; E-mail: sultana.bedoura@dce.butex.edu.bd

Vinitha Moolchand Thadhani (Ed.)

ventional dyeing process utilizes approximately 100,000 distinct dyes, resulting in the generation of around 300,000 tons of wastewater that increases the levels of BOD, COD, TDS, and TSS in the water [1 - 5]. This environmental challenge has led to an urgent demand for alternative dyeing techniques that align with the principles of a circular economy and meet the growing global sustainability objectives.

The emergence of the circular economy highlights the importance of resource efficiency and waste valorization, aiming to mitigate environmental impact while concurrently enhancing the value of waste materials [6, 7]. Due to its polyphenols, tannins, and natural pigments, *Camellia sinensis*, tea, which is one of the most frequently consumed beverages, has the potential to be utilized as a dye. Consequently, it is an untapped resource for producing eco-friendly dyes [8, 9]. Several studies have demonstrated that the dye extracted from tea can serve as an effective colorant for fabric dyeing, enhancing color strength and providing remarkable color fastness against washing, rubbing, and light exposure on cotton, jute, and wool fabric [9 - 13]. Additionally, the extracted dyes improve the biodegradability of the dyed fabric. Several methods, such as Solvent Extraction (SE), Pressurized Extraction (PE), Ultrasound-Assisted Extraction (UAE), Soxhlet Extraction (SoE), Supercritical Fluid Extraction (SFE), and Enzymatic Extraction (EE), are employed to extract dyes from tea waste, including spent tea leaves derived from black, white, oolong, and green tea. These methods use water and ethanol as solvents [14]. Tea waste is the byproduct produced during the manufacturing processes of different tea varieties, such as green, white, black, and oolong. The byproduct comprises the leaves, stalks, and stems that are eliminated during the manufacturing process. Furthermore, spent tea is commonly known as tea waste, which involves drying tea leaves following the steeping process in hot water [15]. Utilized tea bags and leaves fall under the classification of tea waste [16]. Additionally, leaves and stems obtained from the factory or the farmer that are not used in tea production are classified as tea waste [17, 18]. The waste from this tea is utilized for dye extraction through the specified extraction method. So, optimizing various extraction parameters, such as temperature, concentration, pressure, and the material-to-liquor (M: L) ratio, is crucial for enhancing the extracted dye's color fastness and color strength characteristics. Modern statistical optimization approaches, such as Response Surface Methodology (RSM), are used to allow systematic investigation of multiple interacting variables with fewer experiments than traditional methods [19 - 21].

This chapter explores the potential of tea waste, including black, oolong, white, and green, collected from households, factories, and farmers as a natural dye source, positioning it as a viable alternative to synthetic dyes. It begins by discussing the various types of tea and their chemical compositions, highlighting their antimicrobial properties and potential benefits against UV radiation. The chapter then covers several extraction techniques, including aqueous extraction, ultrasound-assisted extraction, and microwave-assisted extraction. Subsequently, optimized parameters for upscaling tea waste (black, green, oolong, and white) into a sustainable dye and its application to various fabrics are examined. Furthermore, various empirical studies and recent developments are discussed, establishing a clear framework for the broader use of dyes derived from tea waste in the textile sector. Additionally, the chapter examines the collection of tea waste from households, factories, and agricultural producers.

TEA WASTE: COMPONENTS AND DYE POTENTIAL

Types of Tea Waste and Their Bioactive Components

Tea waste contains lignocellulosic biomass, including lignin, cellulose, hemicellulose, and catechins (Fig. **1**), along with amino acids, carbohydrates, caffeine, and other bioactive compounds (Table **1**) [14, 22]. The composition of tea waste typically matches that of regular tea [23], exhibiting comparable quantities of its constituents [22, 24]. Among various tea types, green tea is particularly rich in catechins, amino acids, and carbohydrates. It also contains smaller amounts of lipids, caffeine, and polyphenols [25]. Oolong tea contains 22.7% protein, 17.5% lignin, 17.5% cellulose, and small quantities of purine alkaloids and polyphenols [26, 27]. Black tea is characterized by its significant concentration of polyphenolic compounds, catechins, and caffeine. Black tea contains hot water-soluble polysaccharides and proteins, and hot water-insoluble proteins [28, 29].

Table 1. Chemical composition of various types of tea [23].

Types of Tea	Components	% (w/w) of Dry Biomass	Refs.
Green tea	• Catechins • Amino acids • Carbohydrates • Other polyphenols • Caffeine and others • Lipid	30 12 11 4 3 3	[25]
Oolong tea	• Protein • Lignin • Cellulose • Hemicellulose • Purine alkaloids • Polyphenols	22.7 19.5 17.5 16.4	[26, 27]

(Table 1) cont.....

Types of Tea	Components	% (w/w) of Dry Biomass	Refs.
Black tea	• Cellulose • Polyphenols • Catechins • Lignin • Hot water-insoluble proteins • Hot water-soluble polysaccharide and protein • Caffeine	37.2 30-40 10-20 14.7 12.1 9.6 6-8 2.5	[28, 29]

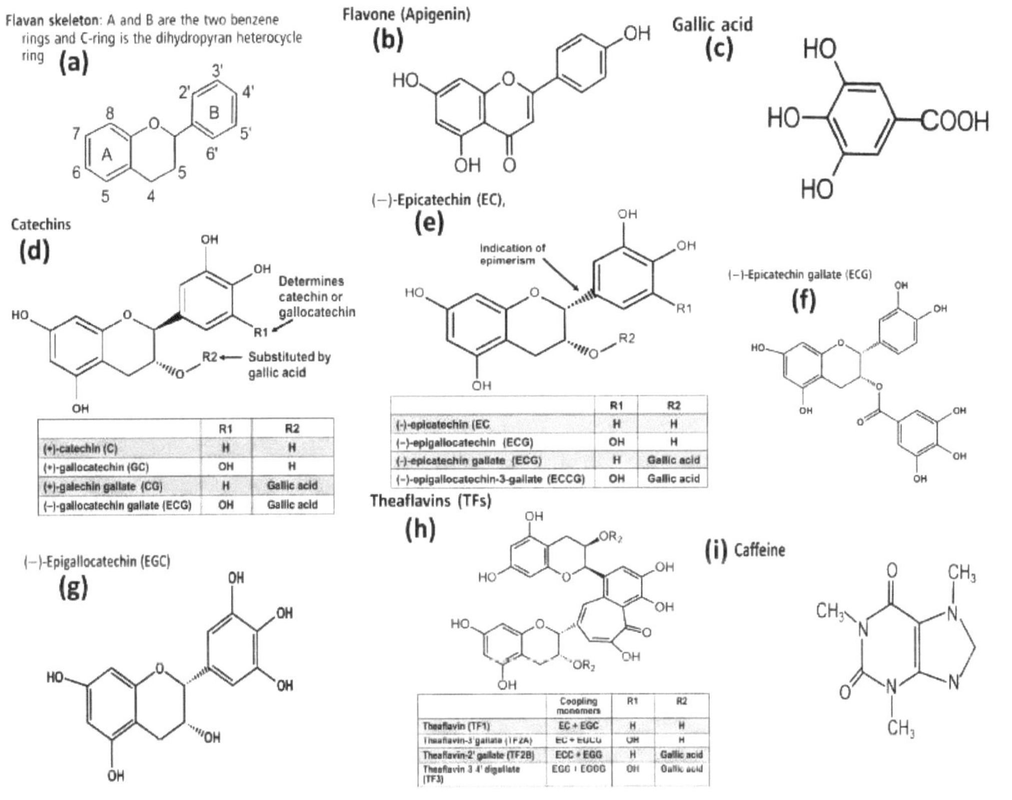

Fig. (1). Chemical structure of bio-active components of tea (a) flavan skeleton, (b) flavone, (c) gallic acid, (d) catechins, (e) (-)-epicatechin (EC), (f) (-)-epicatechin gallate (ECG), (g) (-)-epigallocatechin (EGC), (h) theaflavins (TFs), (i) caffeine [31].

The bioactive constituents found in both tea and its waste exhibit notable similarities. Tea consists of more intricate chemical structures throughout the processing stages [30, 31]. Within the variety of chemical constituents in tea, catechins emerge as the most significant components. Tea leaves contain eight distinct catechin fractions, with four primary catechins being (-)-epigallocatechin (EGC), (-)-epigallocatechin-3-gallate (EGCG), (-)-epicatechin (EC), and (-)

epicatechin-3-gallate (ECG) [32]. Collectively, these four catechins account for 90% of the total catechin content. Furthermore, tea leaves comprise a variety of components, including polyphenols (flavonoids), amino acids, enzymes, pigments, carbohydrates, alkaloids, methylxanthines, vitamins, minerals, and numerous volatile aromatic compounds, as shown in Fig. (**1**), which contributes to the appealing appearance, aroma, flavor, and overall taste of the teas [33].

Antimicrobial and UV Protection Potential

Both tea and tea waste contain similar polyphenolic compounds, including catechins, theaflavins, and thearubigins [16, 30]. Consequently, tea waste demonstrates significant antioxidant and anti-inflammatory properties, as shown in Fig. (**2**), and the mechanisms of these properties are similar. Due to the polyphenolic compounds, the function of the epithelial barrier is enhanced, improving gut microbiota dysbiosis. The polyphenolic compounds function as direct antioxidant agents by neutralizing Reactive Oxygen Species (ROS), such as peroxyl, hydroxyl, superoxide radicals, singlet oxygen, nitrogen dioxide, nitric acid, and peroxynitrite. This action helps to mitigate oxidative damage to lipids, proteins, and nucleic acids. Furthermore, epigallocatechin gallate (EGCG), which is derived from catechins, can neutralize a range of Reactive Oxygen Species (ROS). Catechins possess vicinal dihydroxy and trihydroxy configurations, which facilitate the chelation of metal ions. Consequently, the formation of free radicals is inhibited, promoting electron delocalization that improves their efficiency in quenching radicals [34]. Green tea, white tea, and black tea exhibit antioxidant potential due to their phenolic and polyphenolic components, which demonstrate antioxidant and anti-inflammatory characteristics (Table **2**) [35 - 37]. Ma *et al.* provided evidence that Xinyang white tea exhibits remarkable antioxidant properties, which are attributed to its composition of epicatechin (EC), epigallocatechin (EGC), epicatechin gallate (ECG), and epigallocatechin gallate (EGCG). However, the tea extract's chemical compounds and antioxidant potential may vary depending on the harvest season [37].

Table 2. Summary of antioxidant properties of tea extract.

Tea	Components*	Properties	Refs.
Xinyang white tea	EC, EGC, ECG, EGCG	Antioxidant	[37]
Green tea	Phenolic compounds	Antioxidant	[36]
Black tea waste	-	Antioxidant	[35]
Black tea	C, EC, EGC, ECG, EGCG, TF	Antioxidant, anti-inflammatory	[38]
Tea leaves	C, GC, EC, EGC, ECG, CG, GCG, EGCG	Antioxidant	[39]

(Table 2) cont.....

Tea	Components*	Properties	Refs.
Zijuan tea	EGC, EC, ECG, EGCG, C, CG, GC, and GCG	Antioxidant	[40]
***Catechin (C), gallocatechin (GC), epicatechin (EC), epigallocatechin (EGC), epicatechin gallate (ECG), Catechin Gallate (CG), gallocatechin gallate (GCG), and epigallocatechin gallate (EGCG), Theaflavins (TF).**			

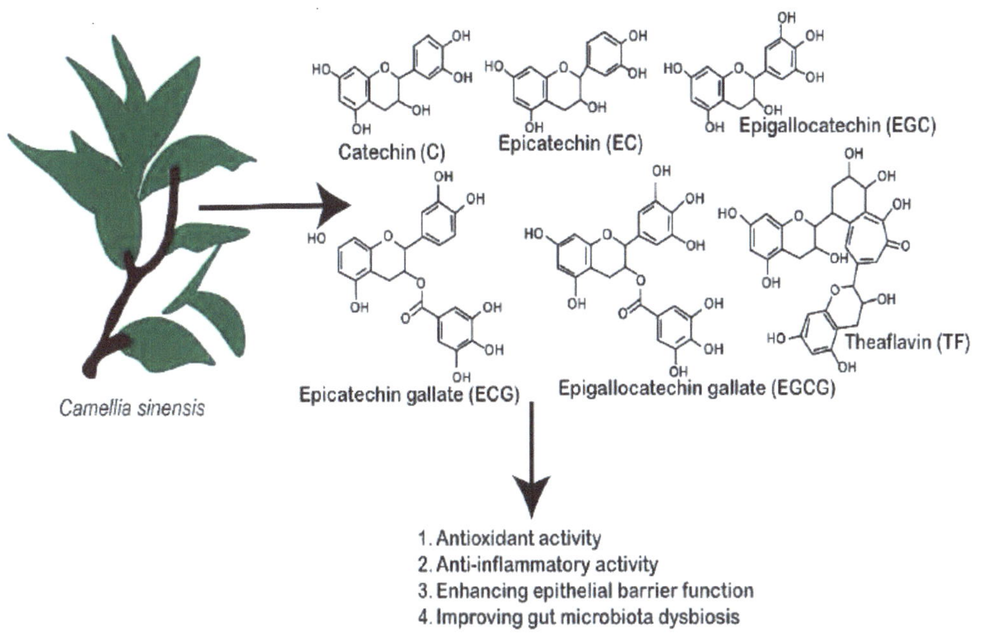

Fig. (2). The structure of the main tea polyphenols and their role in antioxidant and anti-inflammatory activities [16, 30].

Polyphenolic compounds in tea waste contribute to its remarkable UV-resistant properties when chitosan and melamine are added to the extract (Table **3**). However, tea's UV protection properties do not directly correlate with its antioxidant potential. Bonet-Aracil *et al.* employed green, black, and red tea extracts as dyes to color the cotton fabric. The dyed fabric exhibited UV-resistant characteristics; however, no correlation was found between its antioxidant potential and UV protection properties [41]. Khan *et al.* have demonstrated that multifunctional films incorporating chitosan (Ch), gelatin (Gel), green tea, and carbon dots exhibit remarkable UV protection capabilities. The Ch/Gel film containing 3 wt% GT-CD demonstrated significant antioxidant and antimicrobial properties. The multifunctional film demonstrated a remarkable capacity to block 90.23% of UV-A and 99.65% of UV-B light, highlighting the UV-protection potential inherent in tea extract [42].

Table 3. Summary of UV-protection properties of dye extracted from tea.

Tea	Chemicals	Characteristics	Refs.
Tea	Chitosan	• Higher UV protection ability than green and black tea	[41]
Tea	Tea polyphenol-melamine-phenylphosphonic acid	• UPF value: 35.2	[43]
Tea	Tea polyphenol lignin	• Improve UV-resistant properties	[44]
Tea	Tea polyphenol/citric acid	• Exhibit UV-resistant properties	[45]
Vine tea	Vine tea- Musa basjoo- Silk cross-linking	• UPF value: 173.58	-
Green tea	Chitosan/gelatin-based multifunctional film with green tea carbon dots	• Exhibit excellent UV-blocking properties (□90.23% of UV-A, 99.65% of UV-B)	[42]

EXTRACTION OF DYE FROM TEA WASTE

Collection of Tea Waste

Tea waste—originating from oolong, green, white, and black tea—can be sourced through various means, including manufacturing waste, tea gardens, and post-consumer sources. Factory Tea Waste (FTW) refers to discarded or unutilized tea leaves that are unsuitable for further processing during tea production. These materials, often comprising leaves and stems, are commonly used for dye extraction [18].

In addition, raw tea leaves collected directly from tea garden farmers, typically those not meeting commercial standards, can serve as another viable source. Spent tea, the residue left after brewing, is also a widely utilized form of tea waste. It can be collected from households, local tea stalls, and tea shops.

Pretreatment of Tea Waste

A proper pretreatment process is essential for effective dye extraction from tea waste, especially spent tea, as it significantly improves dye quality. Regardless of the source—factory waste, farm discards, or post-brewing residues—all collected tea waste undergoes a cleansing procedure to remove impurities, dust, and organic residues.

The waste material includes discarded leaves and stems from various tea types (oolong, black, green, white) and manufacturing classifications such as Broken Mixed Fannings (BMF) and tenders [46, 47]. After collecting, the tea waste is thoroughly rinsed with clean water and then subjected to a drying phase at ambient temperature. Drying is typically carried out under sunlight for several days to ensure complete moisture removal. Once dried, the samples are carefully stored for subsequent dye extraction procedures [16].

Extraction Techniques

Solvent Extraction (SE)

The solvent extraction method involves using water and ethanol as solvents to obtain dye from tea waste [13]. The extraction process is conducted at a boiling temperature within a specified medium. This method is both environmentally sustainable and economically viable, as it employs a reduced quantity of toxic chemicals during the extraction process. The tea waste was initially cleaned and dried at ambient temperature to eliminate moisture content. Subsequently, the tea waste underwent mechanical milling to reduce its particle size. After milling, water was introduced at 100 ºC for 2 hours. The extraction process utilizes a material-to-liquor ratio of 1:10. The solution was ultimately obtained through filtration. The dye obtained from this process had 14.5% water and 12.1% methanol, with an acidic pH [18]. Tanvir *et al.* utilized a similar extraction technique to obtain dye from discarded tea bags, which were initially dried at room temperature. After drying, water was added to maintain a material-to-liquor ratio of 1:30. The extraction process was conducted for 30 minutes at the boiling point. The pH of the solvent-extracted solution was carefully kept at 5.17±0.5. Continuous stirring was implemented to enhance the efficiency of the extraction method [9].

Most solvent extractions are performed at boiling temperatures [11, 12, 41, 50]. Bonet-Aracil *et al.* showed that the extraction can be conducted at low temperatures. Tea waste was first crushed manually with a mortar to decrease particle size. Dry ice was incorporated into the mortar. The powder was subsequently added to the ethanol-water mixture at a volume ratio of 70:30. The solution was mixed for 1 minute. Following this, the solution was homogenized for one minute using a Polytron. The solutions were centrifuged at 2500 rpm and 4 °C for 5 minutes, and the resulting supernatant was filtered through filter paper. This fluid was then used as a dye for dyeing [41]. The aqueous extraction method presents certain drawbacks. The dye obtained from tea waste exhibits reduced color strength. Metallic mordants are used to enhance the color-fastness properties of the dye. It is also a time-consuming process compared to other dyeing methods, indicating that this dyeing method is less efficient than others. Table **4** summarizes the aqueous extraction of tea dye for textile dyeing.

Table 4. Summary of dye extraction from tea waste using the SE method.

Solvent	Time (hour)	pH	Temperature (°C)	Material to liquor ratio (M: L)	Refs.
Water	2	-	100	1:10	[18]
Water	0.5	5.17±0.5	100	1: 33	[9]
Water	1	-	100	1: 20	[48]
Water	0.5	-	100	1: 20	[49]
Water	2	-	100	1: 50	[41]
Ethanol-water	-	-	4	1: 50	[41]
Water	0.67	Basic	100	-	[50]
Water	2-3	-	90-100	-	[51]
Water	0.95	-	87	1:30	[17]
Water	1.25	-	100	1: 5	[52]
Water	2	-	5, 70, 100	1: 20	[11]
Water	-	7	100	1: 4	[53]
Sodium carbonate	1	4	-	-	[54]

Pressurized Extraction (PE)

Pressurized Extraction (PE), which is also referred to as Pressurized Liquid Extraction (PLE), Accelerated Solvent Extraction (ASE), or Subcritical Water Extraction (SWE) (depending on the solvent and conditions), is a process that utilizes specific temperature and pressure to optimize the extraction of bioactive compounds from solid plant matrices, such as tea leaves. To enhance the solubility and mass transfer of compounds, such as catechins, caffeine, and polyphenols, this method typically employs water or organic solvents (methanol, ethanol, ethyl lactate) at temperatures ranging from 25–200 °C and pressures that are sufficient to maintain the solvent in a liquid [55 - 58].

Initially, the tea waste is dried out, ground, or pulverized to reduce the particulate size and increase the surface area. Subsequently, the tea leaves/powder are deposited into a stainless steel extraction cell. Organic solvents, such as ethanol, methanol, ethyl lactate, or water (in subcritical conditions), are chosen based on the target compounds. To prevent solvent boiling, the operating conditions encompass a temperature range of 50–200 °C and a pressure range of 10–500 MPa. The time frame for operation is 1–60 minutes. The process flow of pressurized liquid extraction is shown in Fig. (**3**). The solvent is poured into the extraction cell, and the system is sealed. The solvent enhances the solubility and desorption of analytes by permeating the tea matrix at elevated temperatures and pressures. Flushing the extract into a collection receptacle is a process that can be repeated for multiple cycles to guarantee its completion. The extract is purified as necessary during the post-processing phase, with the solvent being evaporated or concentrated to reduce the extraction time and solvent consumption compared to conventional methods. The solubility and diffusivity of target compounds are enhanced by the elevated temperature and pressure in the PE method, resulting in a more efficient and rapid extraction. PE typically necessitates smaller solvent volumes than Soxhlet extraction. Compounds are extracted in minutes instead of hours, which enhances throughput and minimizes degradation. The method adheres to the principles of green chemistry by employing water as a solvent, thereby avoiding toxic organic solvents [55 - 58]. This extraction method was used by L. Yang *et al.* on waste tea leaves collected from a tea garden [59].

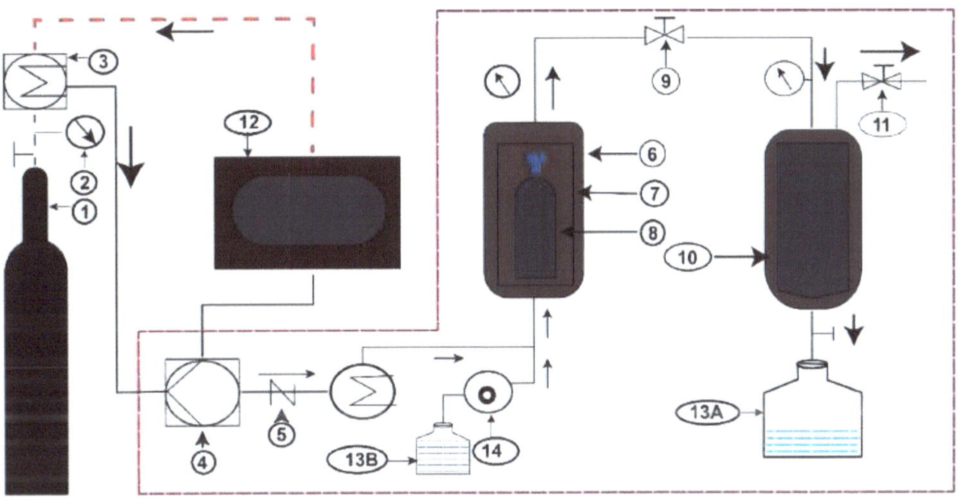

Fig. (3). Process flow of the pressurized liquid extraction unit [56].

However, the PE procedure necessitates acquiring and maintaining specialized, high-pressure extraction systems, which can be expensive. Despite the efficiency of the PE method, the high temperatures may degrade heat-sensitive compounds, such as specific catechins or aroma compounds in tea. Undesired components, including proteins or chlorophylls, may be co-extracted under high temperatures. The extraction parameters (temperature, pressure, solvent type, and time) must be meticulously optimized to balance yield, selectivity, and compound stability, as shown in Table **5**.

Table 5. Optimized parameters of dye extraction from tea waste using the PE method.

Tea	Solvent	Time (min)	Pressure (MPa)	Temperature (°C)	Flowrate	Refs.
Black	• Water • Methanol • Ethanol	-	-	40–80	2 mL/min	[60]
Sweet	• Hot water	40	0.16	113	20 mL/g	[58]
Oolong White	• Water • Methanol • Ethanol	5-10	200-500	100–200	-	[61]
Zhongcha	• Water	60	0.1	120	-	[62]
Black	• Water • Ethanol	-	5	160	12 mL/min	[63]
Green	• Water • Ethanol	1	500	RT*	-	[64]
Green	• Water • Ethanol	15	400	RT*	-	[65]
Green	• Water • Methanol • Ethanol	20	50-200 bar	RT*	0.5 mL/min	[56]
Green	• Ethyl lactate	20	10	100	-	[66]
Green	• Water • Acetone • Methanol • Ethanol	1	500	RT*	-	[67]
Green	• Water • Ethanol	15	498.8	RT*	20.8 mL/g	[55]
***RT: Room temperature**		-	-	-	-	-

Ultrasound-Assisted Extraction (UAE)

Ultrasound-assisted extraction is an effective and environmentally friendly technique for extracting dye from tea waste (FTW [18], BMF [46], stems [47]). To break down the cell walls and release the dye from the materials into the solvent, high-frequency sound waves are used [68]. This method enhances the dyeing efficiency, minimizes extraction time, and decreases solvent usage [16, 68]. The maximum extraction efficiency is observed at 20 kHz and 50 °C heating [68].

The ultrasound-assisted extraction method collected tea waste, washed it, and dried it at room temperature. This process was conducted over 2 days in sunlight, followed by an oven-drying phase for 30 minutes at a temperature of 105°C. The particle size was diminished by employing a milling machine to crush the tea waste. The sample's surface area in contact with the solvent was consequently increased. The ultrasound-assisted extraction process was conducted in a water bath, with the solution being sonicated at a frequency of 20–25 kHz for 30 minutes at a temperature of 60 °C. The optimization of extraction parameters was conducted to achieve optimal results [69]. However, optimizing extraction parameters—including ultrasonic frequency, power, extraction time, and temperature—plays a crucial role in determining the efficiency of the extraction process. The parameters vary based on the specific bioactive compounds being extracted and the type of tea waste used, necessitating comprehensive experimentation to determine the optimal conditions [70].

Microwave-Assisted Extraction (MAE)

Microwave-Assisted Extraction, also known as MAE, is a highly effective extraction technique that utilizes microwave energy to increase the temperature of the solvent in contact with the sample, thereby facilitating the extraction of materials from the target sample [71, 72]. This approach utilizes electromagnetic radiation based on the fundamental principle of thermal energy conversion. This extraction method is notably efficient, significantly reducing the time required. The efficient and consistent heating achieved through microwaves reduces energy usage, contributing to a more sustainable process [10]. MAE can potentially improve the quality of the extracted dye [29].

The tea waste was first dried at room temperature, followed by grinding to allow it to settle. The solvent, ethanol, was subsequently introduced to the powder. 450 W of microwave energy was employed for the effective extraction process, as shown in Fig. (**4**). Subsequently, the solution was allowed to settle for one hour before undergoing filtration. The specimen underwent filtration and was subsequently extracted using a rotary vacuum evaporator at a temperature range of 50-60°C for 45 minutes [29]. The efficacy of the MAC method depends on several factors, including microwave power, extraction technique, and the type of solvent used. Enhancing these factors is essential for achieving the desired results, a process that may be time-consuming [29].

Modern statistical techniques, such as RSM, can be integrated with MAE to optimize these extraction parameters, like microwave power, solvent type, extraction time, and material-to-liquid ratio [19, 20].

The advantages and limitations of SE, UAE, and MAE are shown in Table **6.**

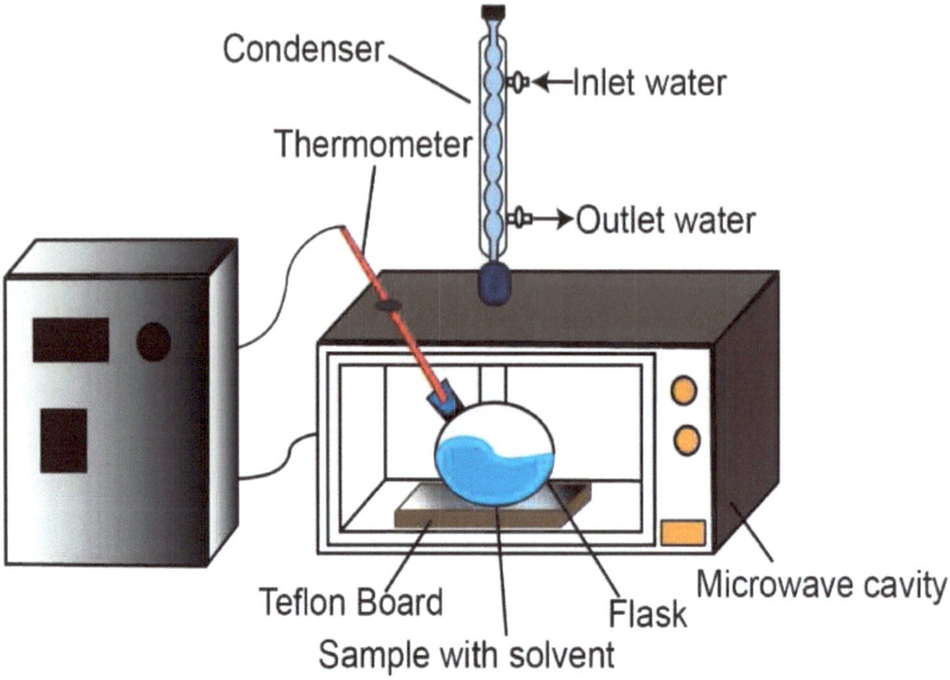

Fig. (4). Schematic diagram of the MAE system [73].

Table 6. Advantages and limitations of SE, UAE, and MAE [74].

Extraction	Advantages	Limitations
SE	• Low installment cost • Purified extraction • A wide range of solvents o Water o Methanol o Ethanol o Ethyl acetate o Hexane	• Required o Batch processing o Post-processing o High amounts of solvent o Additional purification in post-processing o High temperature • Risk of high degradation of compounds
UAE	• Exposure time: Less • Recovery: Maximum • Solvent use: Less	• Required o Batch processing o Post-processing
MAE	• Set-up: Simple • Scale-up: Easy • No post-processing • Temperature: Low	• Extraction time: High • Wave distribution: Not uniform

Irradiation Extraction (IE)

Far-infrared irradiation (FIR) has been applied to improve the extractability of polyphenols and flavanols from green tea leaves (*Camellia sinensis*), making it a promising technique for dye extraction from tea waste [75]. FIR heating—applied during roasting, drying, or as a post-drying treatment—was shown to significantly enhance the chemical constituents responsible for coloration. For example, FIR treatment increased the total phenolic content of processed green tea from 116.30 to 171.77 mg/g and flavanol content from 17.54 to 24.76 mg/g, with corresponding increases in epicatechin gallate (2.41 → 4.59 mg/mL) and epigallocatechin gallate (20.61 → 28.54 mg/mL) compared with non-irradiated control samples [75]. FIR irradiation at 90 °C significantly increased nitrite scavenging activity, while the levels of epigallocatechin and epigallocatechin gallate continued to rise up to 110 °C [76]. In another study, FIR irradiation at 90 °C for 10 min increased total phenols from 244.7 to 368.5 mg/g and flavanols from 122.0 to 178.7 mg/g, while maintaining minimal color change in the final extract. These enhancements are attributed to applying FIR irradiation during the roasting step, before extracting the green tea leaves in water [77].

In a typical FIR-assisted extraction workflow, tea waste is washed, dried to 3–5% moisture, milled, and subjected to FIR heating at 80–150 °C for 10 minutes, either replacing or supplementing conventional roasting/drying steps. Extracts are then obtained by soaking 1 gram of treated leaves in 100 mL hot water (70–75 °C) for 10 minutes [75]. Optimization of FIR temperature and timing is crucial, as excessive heating can degrade catechins, while controlled irradiation increases pigment release and enhances the potential dye yield. Overall, FIR irradiation can significantly improve phenolic availability and therefore may enhance the color strength and efficiency of tea-waste-based dyes.

Gamma irradiation (5–20 kGy) has been explored as a processing technique to modify the chemical profile of green tea extracts and improve color characteristics, presenting another potential method for enhancing dye extraction from tea waste. Irradiation of dried green tea leaves in 70% ethanol resulted in a distinct increase in lightness (L-value) and a decrease in a* and b* values, shifting dark brown extracts toward a brighter yellow appearance—an effect beneficial for applications requiring cleaner, less turbid natural dyes [78]. Additional studies using 5 kGy showed that phenolic and tannin contents of various Bulgarian herbal teas were largely maintained after irradiation, with some samples (*e.g.*, Staroplaninski tea) exhibiting higher antioxidant potential after treatment [79].

For dye extraction, gamma irradiation helps disrupt cellular structures and alter pigment–protein complexes, improving solubility and modifying extract color without substantial degradation of functional compounds. Parameter optimization—primarily irradiation dose and solvent system—is essential, as higher doses may reduce specific phenolics while preserving or enhancing others. RSM can be utilized to optimize the process parameters [20]. Overall, gamma irradiation offers a controlled method to lighten tea extracts and potentially improve their applicability as natural textile dyes.

Soxhlet Extraction (SoE)

A conventional extraction technique, where continuous solvent refluxing occurs to extract various components, including phenolic and flavonoid compounds from tea leaves, is called the Soxhlet extraction process (SoE).

Waste tea leaves are initially dried. After that, the particulate size is reduced, and the surface area is increased by milling the desiccated tea leaves. After milling, the tea leaves have an average particle size of 0.34 mm. Then, the milled tea leaves are inserted into a porous extraction receptacle compatible with the Soxhlet apparatus. SoE employs a solvent capable of dissolving the desired compounds, such as water, ethanol, methanol, dichloromethane, and n-hexane. For this extraction, the Soxhlet apparatus is assembled with the solvent flask and extraction chamber, which includes a condenser and pipette. The solvent in the flask is elevated to its boiling point, which is essential for the SE process's inherent efficacy. As a result, its vapors ascend into the condenser. The target compounds are progressively solubilized as the vapors condense and trickle back into the thimble, percolating through the tea granules. Upon reaching a predetermined level, the solvent that contains the extracted components is sucked back into the simmering vessel. This continuous cycle enables the samples to be repeatedly exposed to new solvents, leading to a comprehensive extraction. To accomplish nearly complete extraction of the soluble compounds, the procedure is typically permitted to continue for 1-3 hours. Rotary evaporation is generally employed to eliminate the solvent and concentrate the extract following extraction. The concentrated extracts are refrigerated for future utilization. Nevertheless, further processing could eliminate the impurities [80 - 82]. The schematic setup of SoE is shown in Fig. (**5**). This method is cost-effective, well-standardized, and straightforward for small-scale applications. The apparatus is comparatively easy to set up, and the SoE extraction method is highly efficient due to continuous solvent recycling [83]. Dębczak *et al.* used this method to extract phenolic components and catechins from tea by optimizing the extraction parameters. The highest color yields, phenolic contents, flavonoid, and gallic acid content are obtained when 40% water/ethanol solutions are used, where the temperature, pressure, and time are 85 °C, 1 bar, and 3 hours, respectively [84].

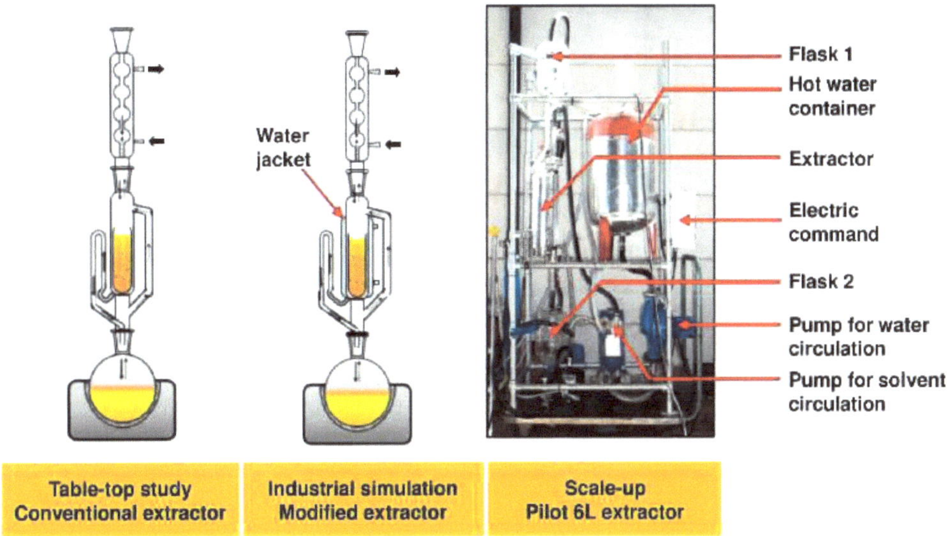

Fig. (5). Soxhlet extraction set-up [88].

However, the SoE procedure can last up to 48 hours and is time-consuming. To maintain the refluxing process for several hours, a substantial quantity of toxic and non-biodegradable solvents is employed [85 - 87].

Supercritical Fluid Extraction (SFE)

Supercritical fluid extraction is an efficient, environment-friendly extraction method where catechins and polyphenols are extracted from tea using supercritical carbon dioxide ($ScCO_2$) at specific temperatures and pressures. This method uses nontoxic, polar solvent ethanol as a co-solvent to increase the extraction efficiency. Extraction yield also increases when ethanol is used [89]. Due to the liquid-like solvation with gas-like diffusion capabilities, this method has an effective extraction medium for extracting catechins and polyphenols from green tea.

The initial stage involves carefully cleaning and drying the collected waste tea leaves. The tea leaves' particle size is then decreased by applying pressure through a laboratory blender. Before extraction, the samples are maintained at a temperature of 4 °C in the refrigerator. The extraction process is conducted under meticulously controlled conditions, as detailed in Table 7, utilizing supercritical carbon dioxide (CO_2, 99.99%) to extract catechins and polyphenols from green tea effectively. The depressurization technique effectively isolates the gas (CO_2) from the samples after the extraction process, carefully cleaning and drying through the separation process, and subsequently storing it in the refrigerator at a temperature of 4 °C [90]. The schematic diagram of SFE extraction is shown in Fig. (**6**). The process allows for the precise adjustment of temperature and pressure, facilitating the efficient extraction of specific molecules and compounds from the tea, such as polyphenols, catechins, and caffeine. This method demonstrates high efficiency, allowing extractions to be completed in 3 hours when utilizing ethanol as a co-solvent. This approach utilizes nontoxic, nonflammable, and recyclable CO_2, categorizing it as an environmentally friendly and efficient method [74].

Table 7. Optimization parameters of catechins and polyphenols extraction from green tea waste [90, 93], respectively, using supercritical fluid.

Tea	Extraction	Parameters					Refs.
		Fluid	Pressure	Temperature	Time	Co-solvent	
Green tea	Catechins	CO_2 (99.99%)	25 MPa	60 °C	3 hours	Ethanol (99.99%)	[93]
Green tea	Polyphenols	CO_2 (99.99%)	23.60 MPa	322.15 K	-	Ethanol (65%) (V/V)	[90]

Still, the components obtained through this method present various drawbacks, such as elevated costs, intricate technical requirements, restricted solubility in the absence of ethanol, and a specific extracted profile. The upfront investment and ongoing maintenance expenses associated with SFE are considerable, necessitating the involvement of skilled personnel and seasoned expertise throughout the extraction process. It is imperative to uphold a stringent safety policy throughout the extraction process. Nonetheless, the resulting tea extracts may occasionally fail to align entirely with the component profiles [91].

Enzymatic Extraction (EE)

To enhance the quality of the tea extract, hydrolytic enzymes are used to break down the cell walls and structural components of tea leaves by releasing bioactive compounds, including polyphenols, catechins, and caffeine, which is called the Enzymatic Extraction (EE) method.

Initially, the particle size of tea waste (leaves) is reduced by crushing them to increase the surface area, facilitating improved enzymatic activities. The particle size is typically between 0.5 and 1 mm. A specific weight ratio combines the desiccated tea leaves with water. Then, the enzyme is introduced to the tea-water solution, where the extraction parameters, such as pH, temperature, time, pressure, and enzyme concentration, are optimized to achieve a more effective extraction. The optimized parameters are shown in Table **8**. The isolates are chilled and filtered after the enzymatic reaction. To concentrate on the targeted extraction of specific components from the tea leaves, EE methods minimize thermal degradation and reduce the use of organic solvents [94]. As demonstrated by Chandini *et al.,* the enzymatic degradation of the cell wall structure in this method results in a more significant release of intracellular compounds, resulting in an enhanced extraction yield. The Extractable-Solid-Yield (ESY) was improved by up to 11.5% using pectinase enzyme. The integrity of heat-sensitive bioactive compounds is preserved by the overall enzymatic process, which operates at moderate temperatures and neutral-to-slightly acidic pH values. The method can reduce the co-extraction of extraneous materials while targeting the release of desired bioactive compounds by selecting specific enzymes. This selectivity frequently results in a purer extract. Enzymatic extraction is more sustainable than numerous solvent-based methods due to its frequent use of water as a solvent and its reduced production of hazardous waste [95].

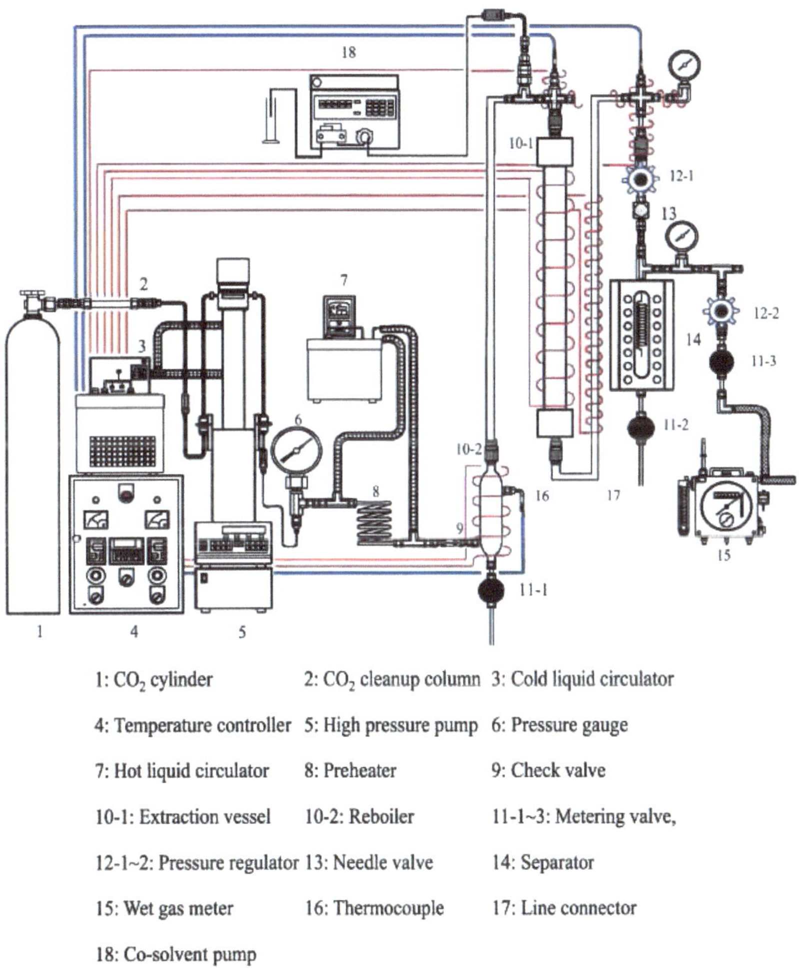

Fig. (6). Schematic diagram of the SFE extraction machine [92].

Table 8. Summary of optimized parameters of polyphenolic components extraction from tea waste by using the enzymatic extraction method.

Enzyme	Source	Optimized parameter	Characteristics	Refs.
Cellulase	*Trichoderma reesei*	• pH: 5.5 • Temperature: 50 °C • Time: 1 hour • Ratio: C: RW-2% (V/W)	• Phenolic content: 85.05 mg • Purity: 25.65%	[96]
• Tannase **• Pectinase**	• *Aspergillus heteromorphous* • *Aspergillus carbonarius*	• pH: 4.8 • Temperature: 40 °C • Time: 2 hours	• Polyphenols recovery: 14. 3% • ESY* improvement: 11.1%	[94]
• Tannase **• Viscozyme**	• *Aspergillus oryzae* • *Aspergillus sp.*	• pH: 4.5-5.5 (T) • Temperature: 35-40 °C (T) • pH: 3.3-5.5 (V) • Temperature: 25-55 °C (V)	• Polyphenol extraction efficiency: Increased	[95]
RW: Ratio of cellulose and raw materials, *ESY: Extractable-solid-yield, T: Tannase, V: Viscozyme				

The high cost of high-purity enzymes necessary for effective extraction frequently increases the production costs on an industrial scale. The extraction efficiency is contingent upon the meticulous regulation of process parameters, such as pH, temperature, enzyme concentration, and reaction duration. Inconsistencies in extraction yield and quality may result from variations. Even though enzymes are selected for their selectivity, there is occasionally a risk of affecting non-target compounds, which could introduce impurities into the final extract. Although laboratory-scale applications have illustrated the advantages of enzymatic extraction, the process can be challenging to scale up due to the sensitivity of enzymatic reactions and the necessity for precision process control [94].

Optimization Parameters

The dye derived from tea extract is significantly influenced by several factors, including the type of tea, the solvent used, pH levels, temperature, duration, and the material-to-liquor ratio (M: L), as shown in Table **5**. It is essential to optimize these factors to achieve the dye's highest color yield and stability. RSM, a statistical optimization tool, can be used to model the combined effects of these variables and identify optimal extraction conditions with fewer experiments [19, 20]. Black [32], green [36], white [37], and oolong teas [32] exhibit varying concentrations of polyphenols and tannins, significantly influencing their dyeing properties. Black tea, rich in theaflavins and thearubigins, produces deep brown hues, while green tea, known for its high levels of catechins, presents more delicate, lighter tones. Oolong tea showcases a spectrum of intermediate hues, whereas white tea offers the most delicate nuances. Consequently, selecting an appropriate type of tea for dye extraction is essential before the dyeing process. The extraction process employs water, ethanol, and sodium carbonate as solvents, which significantly enhance and optimize the extraction of polyphenols. These solvents contribute to the dye's balanced solubility and improve its overall stability. The pH plays a crucial role in determining the efficiency of dye extraction. The extraction process is typically conducted in both acidic and basic environments.

In some cases, the extraction process occurs within a neutral medium, resulting in a moderate level of color intensity. The ideal temperature range for extracting dye from tea waste is between 50 °C and 100 °C. Elevated temperatures enhance the efficiency of extraction processes. The duration required for dye extraction from tea waste ranges from 30 minutes to 2 hours, ensuring optimal yield while maintaining dye stability. The extraction process utilizes a material-to-liquor (M:L) ratio ranging from 1:4 to 1:50. The key extraction parameters for dye recovery from tea waste are summarized in Table **9**.

Table 9. Optimization parameter for dye extracted from tea waste.

Parameters	Effect	Refs.
Solvent type	• Water • Ethanol • Sodium carbonate	[41, 49, 54]
pH	• Acidic • Basic	[9, 50]
Temperature	• 50 °C-100 °C	[11, 17, 53, 68]
Time	• 0.5-2 h	[9, 11, 17, 50, 53, 68]
M:L	• 1:4, 1:5, 1:10, 1:20, 1:50	[9, 11, 17, 50, 53, 68]

POST-EXTRACTION PROCESSING

Post-extraction processing is crucial to ensure the dye's stability, concentration, and effectiveness. The raw tea liquor often contains insoluble components, plant residues, and impurities that can hinder dye uniformity and fabric penetration. Solid-liquid phase separation is typically achieved through methods such as decanting, centrifugation, and filtration. Decanting is often the initial step, where the clarified upper layer of the extract is gently poured off after allowing solids to settle by gravity. This may be followed by centrifugal separation or conventional filtration to remove coarse particulate matter. Further purification is achieved through membrane technologies such as ultrafiltration or reverse osmosis to eliminate finer insoluble particles before application [97, 98].

In some methods, the dye is precipitated from the extract by altering the pH from alkaline to acidic, followed by centrifugation and vacuum drying at low temperature. Alternatively, the tea liquor may be concentrated under vacuum to approximately 30% solids and then spray dried [99]. Evaporation using a water bath is another approach to obtain a semi-dry mass, which can be further dried using freeze-drying or spray drying to yield powdered dye suitable for storage and later use [100]. These post-processing steps not only improve dye quality but also facilitate long-term preservation and scalability for textile applications.

APPLICATION OF TEA WASTE DYE ON FABRICS

The dye derived from tea waste is used in the textile dyeing process for a range of fabrics, including cotton, jute, silk, and wool, as shown in Table **6** [9, 18, 41]. Initially, a metallic mordant is applied at a boiling temperature for one hour. Following the washing and drying processes, the fabrics undergo cationization, after which dyeing is conducted utilizing the extracted liquors in conjunction with a wetting agent. The dyeing procedure is performed for 1 hour at 60 °C. Following the dyeing process, the fabric undergoes a washing phase using detergent, followed by a drying procedure at ambient temperature [9]. Hossain *et al.* dyed the cotton fabric using dye extracted from spent tea bags, where alum was used as a metallic mordant. 3-Chloro-2-hydroxypropyl trimethylammonium chloride (CHPTAC) was used as a cationization agent, and the dyed fabric exhibited excellent color fastness to washing, rubbing, and perspiration [9]. Sarker *et al.* used factory tea waste to extract tea dye and dye the jute fabric. Metallic mordants, such as ferrous sulfate and copper sulfate, are used as mordants. The extracted dye was added to the dye bath, water, and mordant, and heated at 100 °C for 2 hours. The dyeing was performed in an acidic environment, with a material-to-liquor ratio of 1:10. The dyed jute fabric exhibited excellent color fastness to wash and light [18]. A consolidated summary of reported studies on tea-waste dyeing of cotton, jute, and wool fabrics, including mordant selection and fastness outcomes, is provided in Table **10**.

Table 10. Summary of the application of tea waste as dye in several substrates (cotton, jute, wool).

Fabric	Mordant	Result	Refs.
Cotton	Alum	• Excellent color, washing, rubbing, and perspiration fastness	[9]
Jute	Ferrous sulfate, copper sulfate	• Excellent wash and last fastness	[18]
Cotton	-	• Good fastness properties without a chemical mordant	[48]

(Table 10) cont.....

Fabric	Mordant	Result	Refs.
Wool	-	• Good rubbing and wash fastness	[100]
Cotton	-	• Good color strength	[41]

CHALLENGES AND LIMITATIONS

Extracted dye from tea waste has the potential to become an alternative to synthetic dyes, as it is eco-friendly. However, upscaling tea waste to a sustainable textile dye faces several challenges. These include seasonal variability in the composition of tea waste, the environmental impact of mordant, industrial scalability and standardization gaps, and issues with color reproducibility and stability.

Various factors, including geographic origin, processing method, and harvesting season, influence the composition of tea waste. Seasonal variation significantly impacts the concentrations of polyphenols, tannins, and flavonoids, which are essential for the effectiveness of dyeing and the intensity of color [32]. This phenomenon influences the dye formulation, the quality of the extracted dye, and the shade and intensity of the color, thereby determining the optimal dye formulation for fabric coloration. Achieving consistent and long-lasting color from extracted tea dye presents significant challenges due to its inadequate fastness properties, including wash, rubbing, light resistance, and relatively low color strength. The dyeing efficiency of the extracted dye is influenced by several factors, including fabric type, concentration, pH, temperature, and the characteristics of the mordants used [101]. Also, the method of extraction significantly influences the overall efficiency.

Although the small-scale application of tea waste dyeing using metallic mordants has demonstrated promising potential, the industrial-scale implementation encounters various challenges. The absence of standardized methods for extracting and applying dyes derived from tea limits their widespread use in various industries. The extended dyeing duration, elevated dye concentration, and supplementary mordanting required to enhance color strength and fastness properties render tea-extracted dyeing less cost-effective than synthetic dyeing. While the dye derived from tea is considered environmentally friendly, various ecological concerns arise from the mordant used in the natural dyeing process. Nonetheless, mordants are crucial in strengthening the connection between fiber and dye, improving color intensity and durability characteristics. Metallic mordants present significant environmental challenges, primarily due to their toxic nature and potential to contribute to water pollution [102, 103].

CONCLUSION

Enhancing tea extract as a sustainable fabric dye presents a viable solution to address and control pollution within the textile sector and wastewater treatment. Dye extracts from tea waste (FTW, BMF, stems), including spent tea, exhibit comparable colorimetric performance to synthetic dyes while offering significant environmental advantages, including a reduced carbon footprint, decreased wastewater pollution, and enhanced biodegradability. This chapter examines the various extraction methods, potential applications, and utilization of tea extract dye within the textile industry. The aqueous extraction is notable among methods for its effectiveness in industrial-scale applications, utilizing water as the solvent and operating at the boiling point. It is essential to optimize the extraction parameters, such as pH, concentration, temperature, and the material-to-liquid (M:L) ratio, to achieve maximum extraction efficiency. Integrating dye extracts from tea into textile dyeing enables industries to adopt a more circular and environmentally sustainable approach, reducing their dependence on synthetic dyes. Challenges remain, including variability in dye composition due to seasonality and geographic origin, color fastness limitations, the environmental impact of metallic mordants, and difficulties in industrial-scale standardization. Future studies should explore the combined effect of advanced extraction methods, statistical optimization, and irradiation pretreatments to enhance dye yield, stability, and reproducibility. Optimizing parameters in techniques such as solvent-based, microwave-assisted, ultrasound-assisted, enzymatic, and supercritical extractions can enhance dye yield while reducing energy and solvent consumption. Further work is also needed to compare these methods under standardized conditions and to evaluate their industrial feasibility, ensuring that tea-waste dyes can be effectively integrated into eco-friendly textile applications. Emerging irradiation-assisted methods, particularly far-infrared (FIR) and gamma irradiation, offer additional opportunities to enhance pigment release, improve extract brightness, and modify the phenolic profile without substantial loss of functional compounds. Integrating these pretreatments with conventional and optimized extraction techniques may further increase dye yield, improve reproducibility, and expand the applicability of tea-waste dyes in sustainable textile processing.

In summary, extracting dye from tea waste aligns with global sustainability objectives. It represents a significant step forward, including optimization enhancements to color fastness and strength.

ACKNOWLEDGMENTS

The authors wish to recognize the application of AI tools, like Grammarly, during the writing process to enhance the clarity and language of the manuscript. The authors thoroughly reviewed and modified the content as necessary and accept full responsibility for the content of the chapters. Additionally, they want to emphasize that AI was not utilized in their chapters to create or modify images, aid in data analysis, or contribute to their research.

REFERENCES

[1] Kadhom M, Albayati N, Alalwan H, Al-Furaiji M. Removal of dyes by agricultural waste. Sustain Chem Pharm 2020; 16: 100259. [http://dx.doi.org/10.1016/j.scp.2020.100259]

[2] Amalina F, Abd Razak AS, Krishnan S, Zularisam AW, Nasrullah M. A review of eco-sustainable techniques for the removal of Rhodamine B dye utilizing biomass residue adsorbents. Phys Chem Earth Parts ABC 2022; 128: 103267. [http://dx.doi.org/10.1016/j.pce.2022.103267]

[3] Amalina F, Razak ASA, Krishnan S, Zularisam AW, Nasrullah M. Water hyacinth (Eichhornia crassipes) for organic contaminants removal in water – A review. Journal of Hazardous Materials Advances 2022; 7: 100092. [http://dx.doi.org/10.1016/j.hazadv.2022.100092]

[4] Amalina F, Razak ASA, Krishnan S, Zularisam AW, Nasrullah M. Dyes removal from textile wastewater by agricultural waste as an absorbent – A review. Cleaner Waste Systems 2022; 3: 100051. [http://dx.doi.org/10.1016/j.clwas.2022.100051]

[5] Azanaw A, Birlie B, Teshome B, Jemberie M. Textile effluent treatment methods and eco-friendly resolution of textile wastewater. Case Studies in Chemical and Environmental Engineering 2022; 6: 100230. [http://dx.doi.org/10.1016/j.cscee.2022.100230]

[6] Chen X, Memon HA, Wang Y, et al. Circular Economy and Sustainability of the Clothing and Textile Industry. Mater Circ Econ 2021; 3: 12. [http://dx.doi.org/10.1007/s42824-021-00026-2]

[7] Chowdhury NR, Paul SK, Sarker T, Shi Y. Implementing smart waste management system for a sustainable circular economy in the textile industry. Int J Prod Econ 2023; 262: 108876. [http://dx.doi.org/10.1016/j.ijpe.2023.108876]

[8] Zhang L, Santos JS, Cruz TM, *et al.* Multivariate effects of Chinese keemun black tea grades (*Camellia sinensis* var. sinensis) on the phenolic composition, antioxidant, antihemolytic and cytotoxic/cytoprotection activities. Food Res Int 2019; 125: 108516. [http://dx.doi.org/10.1016/j.foodres.2019.108516] [PMID: 31554085]

[9] Hossain T, Das SC, Akhtarujjaman M, Uddin MA, Bedoura S. Eco-friendly dyeing of cotton fabric using used tea bag extracts: A comprehensive study on pH, fabric cationisation, and computational analysis. Current Research in Green and Sustainable Chemistry 2023; 7: 100387. [http://dx.doi.org/10.1016/j.crgsc.2023.100387]

[10] Adeel S, Azeem M, Habib N, *et al.* Sustainable Application of Microwave Assisted Extracted Tea Based Tannin Natural Dye for Chemical and Bio-Mordanted Wool Fabric. J Nat Fibers 2023; 20(1): 2136322. [http://dx.doi.org/10.1080/15440478.2022.2136322]

[11] Lambrecht L, Gisbert-Payá J, Bou-Belda E, Bonet MÁ. Optimization of tea extracts composition to

dye cotton. Time and temperature influence. Journal of Applied Research in Technology & Engineering 2020; 1(1): 3-7.
[http://dx.doi.org/10.4995/jarte.2020.13480]

[12] Rehman A, Rehman M, Naveed M, Javed K, Khan A. Eco-Friendly Dyeing of Cotton with Natural Colorants Using Natural Mordants Obtained from Aloe Vera. J Indian Chem Soc 2024; 101(8): 101182.
[http://dx.doi.org/10.1016/j.jics.2024.101182]

[13] Yılmaz F, Bahtiyari Mİ. Use of tea and tobacco industrial wastes in dyeing and antibacterial finishing of cotton fabrics. AATCC Journal of Research 2020; 7(5): 25-31.
[http://dx.doi.org/10.14504/ajr.7.5.4]

[14] Pompei F, Santanatoglia A, Fioretti L, Vittori S, Mannozzi C. Impact of different brewing systems and extraction technologies on the antioxidant profile of black spent tea: A comprehensive analysis. Food and Bioprocess Technology 2025; 18(7): 6650-63.
[http://dx.doi.org/10.1007/s11947-025-03840-7]

[15] Jezerska L, Sassmanova V, Prokes R, Gelnar D, Peikertova P. Spent tea leaves and tea bags - Promising biofuels? Renew Energy 2025; 238: 121841.
[http://dx.doi.org/10.1016/j.renene.2024.121841]

[16] Çakmak TG, Saricaoglu B, Ozkan G, Tomas M, Capanoglu E. Valorization of tea waste: Composition, bioactivity, extraction methods, and utilization. Food Sci Nutr 2024; 12(5): 3112-24.
[http://dx.doi.org/10.1002/fsn3.4011] [PMID: 38726441]

[17] Xu H, Yang C, Song H-y. Eco-Friendly Dyeing and Functional Finishing of Organic Cotton Using Optimized Oolong Tea Stems (Agricultural Waste) Through Response Surface Methodology. Molecules 2025; 30(3): 509.
[http://dx.doi.org/10.3390/molecules30030509]

[18] Sarker MKU, Haque MM, Hasan MR, Sultana S, Ray SK, Shaikh MAA. Utilization of factory tea (*Camellia sinensis*) wastes in eco-friendly dyeing of jute packaging fabrics. Heliyon 2024; 10(10): e30948.
[http://dx.doi.org/10.1016/j.heliyon.2024.e30948] [PMID: 38779026]

[19] Aftab M, Ahmad T, Adeel S, Javed M. Microwave Assisted Natural Dye Extraction from Amaltas (*Cassia fistula*) for Cotton Dyeing: Process Optimization Using Modern Statistical and Machine Learning Algorithms. Nat Prod Commun 2025; 20(6): 1934578X251355408.
[http://dx.doi.org/10.1177/1934578X251355408]

[20] Hihat S, Touati N, Sellal A, Madani K. Response Surface Methodology: An Optimal Design for Maximising the Efficiency of Microwave-Assisted Extraction of Total Phenolic Compounds from Coriandrum sativum Leaves. Processes (Basel) 2024; 12(5): 1031.
[http://dx.doi.org/10.3390/pr12051031]

[21] Zhou F, Kaewboucha M, Apiwathnasorn C. Supercritical CO_2 Extraction of Phoenix Dancong Tea Oil: Process Optimization and Fragrance Retention on Textiles. Processes (Basel) 2025; 13(11): 3503.
[http://dx.doi.org/10.3390/pr13113503]

[22] Barathi M, Kumar ASK, Kodali J, Mittal S, Samhith GD, Rajesh N. Probing the Interaction between Fluoride and the Polysaccharides in Al(III)- and Zr (IV)-Modified Tea Waste by Using Diverse Analytical Characterization Techniques. ChemistrySelect 2017; 2(31): 10123-35.
[http://dx.doi.org/10.1002/slct.201701774]

[23] Kiyama R. Estrogenic biological activity and underlying molecular mechanisms of green tea constituents. Trends in Food Science & Technology 2020; 95: 247-260.
[http://dx.doi.org/10.1016/J.TIFS.2019.11.014]

[24] Sui W, Xiao Y, Liu R, Wu T, Zhang M. Steam explosion modification on tea waste to enhance bioactive compounds' extractability and antioxidant capacity of extracts. J Food Eng 2019; 261: 51-9.
[http://dx.doi.org/10.1016/j.jfoodeng.2019.03.015]

[25] Kiyama R. Estrogenic biological activity and underlying molecular mechanisms of green tea constituents. Trends Food Sci Technol 2020; 95: 247-60. [http://dx.doi.org/10.1016/j.tifs.2019.11.014]

[26] Ng KW, Cao ZJ, Chen HB, Zhao ZZ, Zhu L, Yi T. Oolong tea: A critical review of processing methods, chemical composition, health effects, and risk. Crit Rev Food Sci Nutr 2018; 58(17): 2957-80. [http://dx.doi.org/10.1080/10408398.2017.1347556] [PMID: 28678527]

[27] Zhao T, Chen Z, Lin X, Ren Z, Li B, Zhang Y. Preparation and characterization of microcrystalline cellulose (MCC) from tea waste. Carbohydr Polym 2018; 184: 164-70. [http://dx.doi.org/10.1016/j.carbpol.2017.12.024] [PMID: 29352907]

[28] Khan N, Mukhtar H. Tea polyphenols for health promotion. Life Sci 2007; 81(7): 519-33. [http://dx.doi.org/10.1016/j.lfs.2007.06.011] [PMID: 17655876]

[29] Tartillah BA. Optimisation Catechin Extract on Black Tea Waste (BTW) using Microwave Assisted Extraction (MAE) (Time Extraction and Ethanol Concentration as Parameters BIO Web Conf. vol. 98: 01003. [http://dx.doi.org/10.1051/bioconf/20249801003]

[30] Unyay H, Altay HO, Perendeci NA, Szufa S, Ozdemir F, Angelidaki İ. Valorisation potential of black tea processing wastes for bioactive compounds recovery and renewable energy production. J Environ Chem Eng 2025; 13(3): 117124. [http://dx.doi.org/10.1016/j.jece.2025.117124]

[31] Samanta S. Potential Bioactive Components and Health Promotional Benefits of Tea (*Camellia sinensis*). Journal of the American Nutrition Association 2022; 41(1): 65-93. [http://dx.doi.org/10.1080/07315724.2020.1827082] [PMID: 33216711]

[32] Engelhardt UH. Tea chemistry – What do and what don't we know? – A micro review. Food Res Int 2020; 132: 109120. [http://dx.doi.org/10.1016/j.foodres.2020.109120] [PMID: 32331637]

[33] Salman S, Öz G, Felek R, Haznedar A, Turna T, Özdemir F. Effects of fermentation time on phenolic composition, antioxidant and antimicrobial activities of green, oolong, and black teas. Food Biosci 2022; 49: 101884. [http://dx.doi.org/10.1016/j.fbio.2022.101884]

[34] Frei B, Higdon JV. Antioxidant activity of tea polyphenols in vivo: evidence from animal studies. J Nutr 2003; 133(10): 3275S-84S. [http://dx.doi.org/10.1093/jn/133.10.3275S] [PMID: 14519826]

[35] Abraham AM, Alnemari RM, Brüßler J, Keck CM. Improved antioxidant capacity of black tea waste utilizing plantcrystals. Molecules 2021; 26(3): 592. [http://dx.doi.org/10.3390/molecules26030592] [PMID: 33498623]

[36] Teixeira Oliveira J, Machado da Costa F, Gonçalvez da Silva T, *et al.* Green tea and kombucha characterization: Phenolic composition, antioxidant capacity and enzymatic inhibition potential. Food Chem 2023; 408: 135206. [http://dx.doi.org/10.1016/j.foodchem.2022.135206] [PMID: 36528993]

[37] Ma B, Wang J, Zhou B, *et al.* Impact of harvest season on bioactive compounds, amino acids and in vitro antioxidant capacity of white tea through multivariate statistical analysis. Lebensm Wiss Technol 2022; 164: 113655. [http://dx.doi.org/10.1016/j.lwt.2022.113655]

[38] Truong VL, Jeong WS. Antioxidant and anti-inflammatory roles of tea polyphenols in inflammatory bowel diseases. Food Sci Hum Wellness 2022; 11(3): 502-11. [http://dx.doi.org/10.1016/j.fshw.2021.12.008]

[39] Wang Z, Zheng C, Ma C, *et al.* Comparative analysis of chemical constituents and antioxidant activity

in tea-leaves microbial fermentation of seven tea-derived fungi from ripened Pu-erh tea. Lebensm Wiss Technol 2021; 142: 111006.
[http://dx.doi.org/10.1016/j.lwt.2021.111006]

[40] Zou C, Li RY, Chen JX, *et al.* Zijuan tea- based kombucha: Physicochemical, sensorial, and antioxidant profile. Food Chem 2021; 363: 130322.
[http://dx.doi.org/10.1016/j.foodchem.2021.130322] [PMID: 34147900]

[41] Bonet-Aracil MÁ, Díaz-García P, Bou-Belda E, Sebastiá N, Montoro A, Rodrigo R. UV protection from cotton fabrics dyed with different tea extracts. Dyes Pigments 2016; 134: 448-52.
[http://dx.doi.org/10.1016/j.dyepig.2016.07.045]

[42] Khan A, Ezati P, Rhim JW. Chitosan/gelatin-based multifunctional film integrated with green tea carbon dots to extend the shelf life of pork. Food Packag Shelf Life 2023; 37: 101075.
[http://dx.doi.org/10.1016/j.fpsl.2023.101075]

[43] Jiang Q, Li P, Liu Y, Zhu P. Flame retardant cotton fabrics with anti-UV properties based on tea polyphenol-melamine-phenylphosphonic acid. J Colloid Interface Sci 2023; 629(Pt B): 392-403.
[http://dx.doi.org/10.1016/j.jcis.2022.09.084] [PMID: 36166966]

[44] Liu B, Zhang W, Zeng J, *et al.* Acid-catalyzed phenolation of lignin with tea polyphenol: Enhancing uv resistance and oxidation resistance for potential applications. Int J Biol Macromol 2024; 267(Pt 2): 131462.
[http://dx.doi.org/10.1016/j.ijbiomac.2024.131462] [PMID: 38614163]

[45] Shao H, Zhang Y, Pan H, *et al.* Preparation of flexible and UV-blocking films from lignin-containing cellulose incorporated with tea polyphenol/citric acid. Int J Biol Macromol 2022; 207: 917-26.
[http://dx.doi.org/10.1016/j.ijbiomac.2022.03.183] [PMID: 35364193]

[46] Najafi H, Rezaei Laye Z, Sobati MA. Thermal conversion potential of tea stems: Experimental investigation, reaction mechanism, and kinetic modeling. Ind Crops Prod 2024; 207: 117753.
[http://dx.doi.org/10.1016/j.indcrop.2023.117753]

[47] Dalpathadu KAP, Nissanka SP, Jayasinghe CVL. Physico-Chemical Properties of Sri Lankan Instant Black Tea Manufactured Using Broken Mixed Fannings (BMF) as Raw Materials Obtained from Different Estates of Three Elevation Categories. Tropical Agricultural Research 2020; 31(4): 91.
[http://dx.doi.org/10.4038/tar.v31i4.8423]

[48] Ren Y, Fu R, Fang K, *et al.* Dyeing cotton with tea extract based on *in-situ* polymerization: An innovative mechanism of coloring cellulose fibers by industrial crop pigments. Ind Crops Prod 2019; 142: 111863.
[http://dx.doi.org/10.1016/j.indcrop.2019.111863]

[49] Preetha R, Rani NA. Dyeing at home from kitchen waste - tea and coffee residue. ShodhKosh J Vis Per Arts 2023; 4(2): 374-84.
[http://dx.doi.org/10.29121/shodhkosh.v4.i2.2023.519]

[50] Adeel S, Ahmad M, FazalurRehman, *et al.* Eco-friendly dyeing of cotton fabric with waste tea leaves-based tannin natural dye. Glob NEST J 2021; 23(3): 365-9.
[http://dx.doi.org/10.30955/gnj.003456]

[51] Vankar PS. Low temperature optimized dyeing of cotton, wool and silk with extract of Camellia sinensis (tea leaves). Journal of Textile Engineering & Fashion Technology 2017; 2: 1.
[http://dx.doi.org/10.15406/jteft.2017.02.00045]

[52] Triwiswara M, Indrayani L. Utilization of black tea waste as natural batik dyes on cotton and silk. IOP Conf Ser Earth Environ Sci 2020; 456(1): 012051.
[http://dx.doi.org/10.1088/1755-1315/456/1/012051]

[53] Zhao Y, Liu J. Extraction of Tea Natural Dye and its Dyeing Properties on Cotton Fabrics. AATCC Journal of Research 2021; 8(2_suppl) (_Suppl.): 86-90.
[http://dx.doi.org/10.14504/ajr.8.S2.17]

[54] Cheng TH, Liu ZJ, Yang JY, Huang YZ, Tang RC, Qiao YF. Extraction of Functional Dyes from Tea Stem Waste in Alkaline Medium and Their Application for Simultaneous Coloration and Flame Retardant and Bioactive Functionalization of Silk. ACS Sustain Chem& Eng 2019; 7(22): 18405-13. [http://dx.doi.org/10.1021/acssuschemeng.9b04094]

[55] Xi J, Xue Y, Xu Y, Shen Y. Artificial neural network modeling and optimization of ultrahigh pressure extraction of green tea polyphenols. Food Chem 2013; 141(1): 320-6. [http://dx.doi.org/10.1016/j.foodchem.2013.02.084] [PMID: 23768364]

[56] Jaiswal SG, Naik S. Pressurized Liquid Extraction of Antioxidant Compounds from Green Tea. Am J Food Technol 2017; 12(6): 358-66. [http://dx.doi.org/10.3923/ajft.2017.358.366]

[57] Yang L, Dai X, Meng Y, Zhou L, Yue X, Comparative Study on Dyed Fabrics of Tea and Tea Stem Based on Kansei Engineering. Sustainability 2023; 15(2): 1134. [http://dx.doi.org/10.3390/su15021134]

[58] Lei J, Li W, Fu MX, *et al.* Pressurized hot water extraction, structural properties, biological effects, and in vitro microbial fermentation characteristics of sweet tea polysaccharide. Int J Biol Macromol 2022; 222(Pt B): 3215-28. [http://dx.doi.org/10.1016/j.ijbiomac.2022.10.094] [PMID: 36243163]

[59] L. Yang, X. Dai, Y. Meng, L. Zhou, and X. Yue, "Comparative Study on Dyed Fabrics of Tea and Tea Stem Based on Kansei Engineering," Sustainability 2023, Vol. 15, Page 1134, vol. 15, no. 2, p. 1134, Jan. 2023. [http://dx.doi.org/10.3390/su15021134]

[60] Souza MC, Santos MP, Sumere BR, *et al.* Isolation of gallic acid, caffeine and flavonols from black tea by on-line coupling of pressurized liquid extraction with an adsorbent for the production of functional bakery products. Lebensm Wiss Technol 2020; 117: 108661. [http://dx.doi.org/10.1016/j.lwt.2019.108661]

[61] Šeremet D, Karlović S, Vojvodić Cebin A, Mandura A, Ježek D, Komes D. Extraction of bioactive compounds from different types of tea by high hydrostatic pressure. J Food Process Preserv 2021; 45(9): e15751. [http://dx.doi.org/10.1111/jfpp.15751]

[62] Gao Y, Zhou Y, Zhang Q, *et al.* Hydrothermal extraction, structural characterization, and inhibition HeLa cells proliferation of functional polysaccharides from Chinese tea Zhongcha 108. J Funct Foods 2017; 39: 1-8. [http://dx.doi.org/10.1016/j.jff.2017.09.057]

[63] He C, Du H, Tan C, *et al.* Semi-continuous pressurized hot water extraction of black tea. J Food Eng 2018; 227: 30-41. [http://dx.doi.org/10.1016/j.jfoodeng.2018.02.001]

[64] Jun X. Caffeine extraction from green tea leaves assisted by high pressure processing. J Food Eng 2009; 94(1): 105-9. [http://dx.doi.org/10.1016/j.jfoodeng.2009.03.003]

[65] Jun X, Shuo Z, Bingbing L, *et al.* Separation of major catechins from green tea by ultrahigh pressure extraction. Int J Pharm 2010; 386(1-2): 229-31. [http://dx.doi.org/10.1016/j.ijpharm.2009.10.035] [PMID: 19874878]

[66] Villanueva Bermejo D, Ibáñez E, Reglero G, Turner C, Fornari T, Rodriguez-Meizoso I. High catechins/low caffeine powder from green tea leaves by pressurized liquid extraction and supercritical antisolvent precipitation. Separ Purif Tech 2015; 148: 49-56. [http://dx.doi.org/10.1016/j.seppur.2015.04.037]

[67] Xi J, Shen D, Zhao S, Lu B, Li Y, Zhang R. Characterization of polyphenols from green tea leaves using a high hydrostatic pressure extraction. Int J Pharm 2009; 382(1-2): 139-43.

[http://dx.doi.org/10.1016/j.ijpharm.2009.08.023] [PMID: 19715745]

[68] Ozsefil IC, Ziylan-Yavas A. Green approach for polyphenol extraction from waste tea biomass: Single and hybrid application of conventional and ultrasound-assisted extraction. Environ Res 2023; 235: 116703.
[http://dx.doi.org/10.1016/j.envres.2023.116703] [PMID: 37474089]

[69] https://www.researchgate.net/publication/338123459_Extraction_of_Natural_Dye_from_Tea_Leaves_and_its_Application_on_Giza_86_Egyptian_Cotton_Fabric

[70] Shen L, Pang S, Zhong M, *et al.* A comprehensive review of ultrasonic assisted extraction (UAE) for bioactive components: Principles, advantages, equipment, and combined technologies. Ultrason Sonochem 2023; 101: 106646.
[http://dx.doi.org/10.1016/j.ultsonch.2023.106646] [PMID: 37862945]

[71] Tsubaki S, Iida H, Sakamoto M, Azuma J. Microwave heating of tea residue yields polysaccharides, polyphenols, and plant biopolyester. J Agric Food Chem 2008; 56(23): 11293-9.
[http://dx.doi.org/10.1021/jf802253s] [PMID: 18998700]

[72] Serdar G, Demir E, Sökmen M. Recycling of Tea Waste: Simple and Effective Separation of Caffeine and Catechins by Microwave Assisted Extraction (MAE). International Journal of Secondary Metabolite 2017; 4(2): 78-89.
[http://dx.doi.org/10.21448/ijsm.288226]

[73] N. Zghaibi, R. Omar, S. M. M. Kamal, D. R. A. Biak, and R. Harun, "Microwave-Assisted Brine Extraction for Enhancement of the Quantity and Quality of Lipid Production from Microalgae Nannochloropsis sp.," Molecules 2019, Vol. 24, Page 3581, vol. 24, no. 19, p. 3581, Oct. 2019.
[http://dx.doi.org/10.3390/molecules24193581]

[74] Pasrija D, Anandharamakrishnan C. Techniques for Extraction of Green Tea Polyphenols: A Review. Food Bioprocess Technol 2015; 8(5): 935-50.
[http://dx.doi.org/10.1007/s11947-015-1479-y]

[75] Park J-H, Lee J-M, Cho Y-J, *et al.* EFFECT OF FAR-INFRARED HEATER ON THE PHYSICOCHEMICAL CHARACTERISTICS OF GREEN TEA DURING PROCESSING. J Food Biochem 2009; 33(2): 149-62.
[http://dx.doi.org/10.1111/j.1745-4514.2008.00203.x]

[76] Lee SC, Kim SY, Jeong SM, Park JH. Effect of far-infrared irradiation on catechins and nitrite scavenging activity of green tea. J Agric Food Chem 2006; 54(2): 399-403.
[http://dx.doi.org/10.1021/jf051866x] [PMID: 16417296]

[77] Kim SY, Jeong SM, Jo SC, Lee SC. Application of far-infrared irradiation in the manufacturing process of green tea. J Agric Food Chem 2006; 54(26): 9943-7.
[http://dx.doi.org/10.1021/jf062524+] [PMID: 17177525]

[78] Jo C, Son JH, Lee HJ, Byun MW. Irradiation application for color removal and purification of green tea leaves extract. Radiat Phys Chem 2003; 66(2): 179-84.
[http://dx.doi.org/10.1016/S0969-806X(02)00273-6]

[79] Janiak MA, Slavova-Kazakova A, Karamać M, *et al.* Effects of Gamma-Irradiation on the Antioxidant Potential of Traditional Bulgarian Teas. Natural Product Communications. 2017; 12(2).
[http://dx.doi.org/10.1177/1934578X1701200210]

[80] Didion Y Patrice, Tjalsma T Gijsbert, Su Z, Malankowska M, Pinelo M. What is next? the greener future of solid liquid extraction of biobased compounds: Novel techniques and solvents overpower traditional ones. Separation and Purification Technology 2023; 320: 124147.
[http://dx.doi.org/10.1016/J.SEPPUR.2023.124147]

[81] Lopes EC, Santos SCR, Pintor AMA, Boaventura RAR, Botelho CMS. Evaluation of a tannin-based coagulant on the decolorization of synthetic effluents. J Environ Chem Eng 2019; 7(3): 103125.
[http://dx.doi.org/10.1016/j.jece.2019.103125]

[82] Cioanca O, Lungu I-I, Mita-Baciu I, *et al.* Extraction and Purification of Catechins from Tea Leaves: An Overview of Methods, Advantages, and Disadvantages. Separations 2024; 11(6): 171. [http://dx.doi.org/10.3390/separations11060171]

[83] P. Kapadia, A. S. Newell, J. Cunningham, M. R. Roberts, and J. G. Hardy, "Extraction of High-Value Chemicals from Plants for Technical and Medical Applications," International Journal of Molecular Sciences 2022, Vol. 23, Page 10334, vol. 23, no. 18, p. 10334, Sep. 2022. [http://dx.doi.org/10.3390/ijms231810334]

[84] Dębczak A, Grzegorczyk A, Świątek Ł, *et al.* Green and sustainable recovery of polyphenols, flavonoids, alkaloids, and pigments from green tea leaves: Comparative analysis of Soxhlet, accelerated solvent, and supercritical fluid extraction techniques. Sustain Chem Pharm 2024; 42: 101810. [http://dx.doi.org/10.1016/j.scp.2024.101810]

[85] S. Raghunath, S. Budaraju, S. M. T. Gharibzahedi, M. Koubaa, S. Roohinejad, and K. Mallikarjunan, "Processing Technologies for the Extraction of Value-Added Bioactive Compounds from Tea," Food Engineering Reviews 2023 15:2, vol. 15, no. 2, pp. 276–308, Feb. 2023. [http://dx.doi.org/10.1007/s12393-023-09338-2]

[86] Patrice Didion Y, Gijsbert Tjalsma T, Su Z, Malankowska M, Pinelo M. What is next? the greener future of solid liquid extraction of biobased compounds: Novel techniques and solvents overpower traditional ones. Separ Purif Tech 2023; 320: 124147. [http://dx.doi.org/10.1016/j.seppur.2023.124147]

[87] I. M. Koina, Y. Sarigiannis, and E. Hapeshi, "Green Extraction Techniques for the Determination of Active Ingredients in Tea: Current State, Challenges, and Future Perspectives," Separations 2023, Vol. 10, Page 121, vol. 10, no. 2, p. 121, Feb. 2023. [http://dx.doi.org/10.3390/separations10020121]

[88] Sicaire A-G, Vian M, Fine F, Joffre F, Carré P, Tostain S, Chemat F. Alternative Bio-Based Solvents for Extraction of Fat and Oils: Solubility Prediction, Global Yield, Extraction Kinetics, Chemical Composition and Cost of Manufacturing. International Journal of Molecular Sciences 2015; 16(4): 8430-53. [http://dx.doi.org/10.3390/ijms16048430]

[89] İçen H, Gürü M. Effect of ethanol content on supercritical carbon dioxide extraction of caffeine from tea stalk and fiber wastes. J Supercrit Fluids 2010; 55(1): 156-60. [http://dx.doi.org/10.1016/j.supflu.2010.07.009]

[90] Wang W, Han S, Zha X, Cheng J, Song J, Jiao Z. Response Surface Optimization of Supercritical Carbon Dioxide Extraction of Tea Polyphenols from Green Tea Scraps. J AOAC Int 2019; 102(2): 451-6. [http://dx.doi.org/10.5740/jaoacint.18-0163] [PMID: 30305199]

[91] da Silva RPFF, Rocha-Santos TAP, Duarte AC. Supercritical fluid extraction of bioactive compounds. Trends Analyt Chem 2016; 76: 40-51. [http://dx.doi.org/10.1016/j.trac.2015.11.013]

[92] Huang KJ, Wu JJ, Chiu YH, Lai CY, Chang CMJ. Designed polar cosolvent-modified supercritical CO_2 removing caffeine from and retaining catechins in green tea powder using response surface methodology. J Agric Food Chem 2007; 55(22): 9014-20. [http://dx.doi.org/10.1021/jf071161q] [PMID: 17914876]

[93] Sökmen M, Demir E, Alomar SY. Optimization of sequential supercritical fluid extraction (SFE) of caffeine and catechins from green tea. J Supercrit Fluids 2018; 133: 171-6. [http://dx.doi.org/10.1016/j.supflu.2017.09.027]

[94] Chandini SK, Rao LJ, Gowthaman MK, Haware DJ, Subramanian R. Enzymatic treatment to improve the quality of black tea extracts. Food Chem 2011; 127(3): 1039-45. [http://dx.doi.org/10.1016/j.foodchem.2011.01.078] [PMID: 25214094]

[95] Hong YH, Jung EY, Park Y, *et al.* Enzymatic improvement in the polyphenol extractability and antioxidant activity of green tea extracts. Biosci Biotechnol Biochem 2013; 77(1): 22-9. [http://dx.doi.org/10.1271/bbb.120373] [PMID: 23291774]

[96] Tc H, Nd N, Lt HA, Ta V, Pv M. Enzyme Assisted Extraction of Polyphenols from the Old Tea Leaves. Journal of Nutrition and Health Sciences 2016; 3(4) [http://dx.doi.org/10.15744/2393-9060.3.404]

[97] Har Bhajan Singh and K. A. Bharati, "Methods of extraction," Handbook of Natural Dyes and Pigments, pp. 9–17, Jan. 2014. [http://dx.doi.org/10.1016/B978-93-80308-54-8.50003-4]

[98] Kumar A, Thakur BK, De S. Selective Extraction of (−)Epigallocatechin Gallate from Green Tea Leaves Using Two-Stage Infusion Coupled with Membrane Separation. Food Bioprocess Technol 2012; 5(6): 2568-77. [http://dx.doi.org/10.1007/s11947-011-0580-0]

[99] Kumar A, Konar A. Dyeing of Textiles with Natural Dyes [Internet]. Natural Dyes. InTech; 2011. [http://dx.doi.org/10.5772/21341]

[100] Ren Y, Gong J, Wang F, *et al.* Effect of dye bath pH on dyeing and functional properties of wool fabric dyed with tea extract. Dyes Pigments 2016; 134: 334-41. [http://dx.doi.org/10.1016/j.dyepig.2016.07.032]

[101] Veysian M, Shams-Nateri A. The Effect of Metal Mordant and Biomordant on Color Strength of Dyed Woolen Yarn with Natural Dyes. Fibers Polym 2023; 24(12): 4337-55. [http://dx.doi.org/10.1007/s12221-023-00375-6]

[102] Dutta P, Mahjebin S, Sufian MA, Razaya Rabbi M, Chowdhury S, Imran IH. Impacts of natural and synthetic mordants on cotton knit fabric dyed with natural dye from onion skin in perspective of eco-friendly textile process. Mater Today Proc 2021; 47: 2633-40. [http://dx.doi.org/10.1016/j.matpr.2021.05.229]

[103] İşmal ÖE, Yıldırım L. Metal mordants and biomordants. Impact and Prospects of Green Chemistry for Textile Technology 2019; (Jan): 57-82. [http://dx.doi.org/10.1016/B978-0-08-102491-1.00003-4]

CHAPTER 6

Method of Application of Waste Tea Dye

Belal Uddin Rabbi[1] and **Sultana Bedoura**[2,*]

[1] *Department of Wet Process Engineering, Bangladesh University of Textiles (BUTEX), Dhaka, Bangladesh*

[2] *Department of Dyes and Chemical Engineering, Bangladesh University of Textiles (BUTEX), Dhaka, Bangladesh*

Abstract: The use of tea waste as a natural dye offers sustainable and eco-friendly methods for dyeing. These methods serve as an effective alternative to synthetic dyes, decreasing toxicity and promoting a circular economy through the reduction of industrial waste. This chapter outlines a structured process for extracting and applying tea-derived dyes to cotton, silk, wool, and flax, covering substrate preparation, dyeing, and post-dyeing treatments. Experimental findings across multiple studies consistently show improved dye uptake, higher color strength, and enhanced wash and rubbing fastness when optimized pretreatments—such as tannic acid treatment, cationization, enzymatic bioscouring, plasma, or UV-assisted modification are applied. Advanced dyeing approaches, including solvent extraction, ultrasound-, microwave-, and plasma-assisted dyeing, further contribute to greater coloration efficiency and reduced chemical demand, despite variations in dyeing conditions among different investigations. In summary, waste tea demonstrates clear, measurable improvements in color performance and environmental impact. This chapter provides a structured pathway for its application and establishes waste tea as a practical and sustainable dye source for future textile processing.

Keywords: Circular economy, Green tea waste, Natural dyeing, Sustainability, Tea waste.

INTRODUCTION

Natural dyes have been utilized since the Stone Age, with many ancient civilizations, including Indian, Egyptian, Roman, and Greek, employing them in various applications [1 - 3]. The synthesis of synthetic dyes in 1856 led to a decline in the use of natural dyes due to the lower cost, greater durability, wider color range, the inability of natural sources to meet the growing dye demand driven by increased textile production, and ease of application associated with

* **Corresponding author Sultana Bedoura:** Department of Dyes and Chemical Engineering, Bangladesh University of Textiles (BUTEX), Dhaka, Bangladesh; E-mail: sultana.bedoura@dce.butex.edu.bd

synthetic dyes [4]. Synthetic dyes bring toxicity and environmental concerns due to the presence of heavy metals and carcinogens, which pose health risks. Consequently, it is banned in several nations [5 - 9]. The utilization of natural dyes extracted from plants, minerals, and animal sources is increased because natural dyes are biodegradable, renewable, and environmentally friendly [10].

Tea (*Camellia sinensis*) serves as a sustainable and natural dye source. The composition primarily includes phenolic compounds like catechins and flavonoids. Within the flavonoid family, therubigins and theaflavins are the main natural dyeing agents for natural fabrics. The primary coloring pigment responsible for the brown color in fabrics is catechin, a naturally occurring pigment found in tea leaves [11 - 13]. The dye is extracted from tea and utilized to dye cellulosic fabric. Microwave irradiation and ultrasound mechanisms are utilized for extraction. Various dyeing methods, including ultrasound and microwave-assisted dyeing, mordant dyeing, pad-stream dyeing, and simultaneous dyeing, are employed, with fabrics undergoing pre- and post-treatment processes [14 - 16].

Tea leaves serve a dual role in the dyeing process, acting both as a natural dye and a mordant, thereby eliminating the need for synthetic or metal-based mordants. This is due to the presence of tannins, which have inherent mordanting properties that enhance dye fixation on fabrics. The natural tannins in tea form strong complexes with textile fibers, improving the uptake and retention of color. As a result, fabrics dyed with tea exhibit enhanced color fastness properties, making them more resistant to washing and light exposure. However, to further improve the durability and depth of the color, heavy metal mordants such as alum, copper sulfate ($CuSO_4$), and ferrous sulfate ($FeSO_4$) are sometimes employed. These mordanting agents create stronger bonds between the dye and the fabric, resulting in richer shades and improved resistance to fading over time [1, 16, 17].

Furthermore, advanced techniques such as cationization, pre- and post-mordanting, pad-stream dyeing, microwave-assisted dyeing, and dip-dyeing have been utilized to enhance the colorimetric properties of tea-dyed fabrics. These methods improve dye uptake, color uniformity, and fastness, ensuring deeper and more stable shades on textile fibers [12, 15, 18, 19].

This chapter examines practical methodologies for applying waste tea-derived dyes to various fabrics, including cotton, wool, silk, and flax. It emphasizes sustainable and efficient techniques suitable for small-scale artisanal use and industrial application. The outline of the method of application of the waste tea dye is shown in Fig. (**1**). The potential of tea waste as a natural dye is recognized; however, standardized protocols for its extraction, application, and fixation are still lacking. This discussion covers various dyeing methods, including fabric preparation and post-treatment techniques to improve dyed fabrics' color fastness and intensity. This chapter connects theoretical knowledge with practical application, providing researchers, textile engineers, and eco-conscious industries with strategies to utilize waste tea as an eco-friendly alternative to synthetic dyes, promoting circular economy principles in material design.

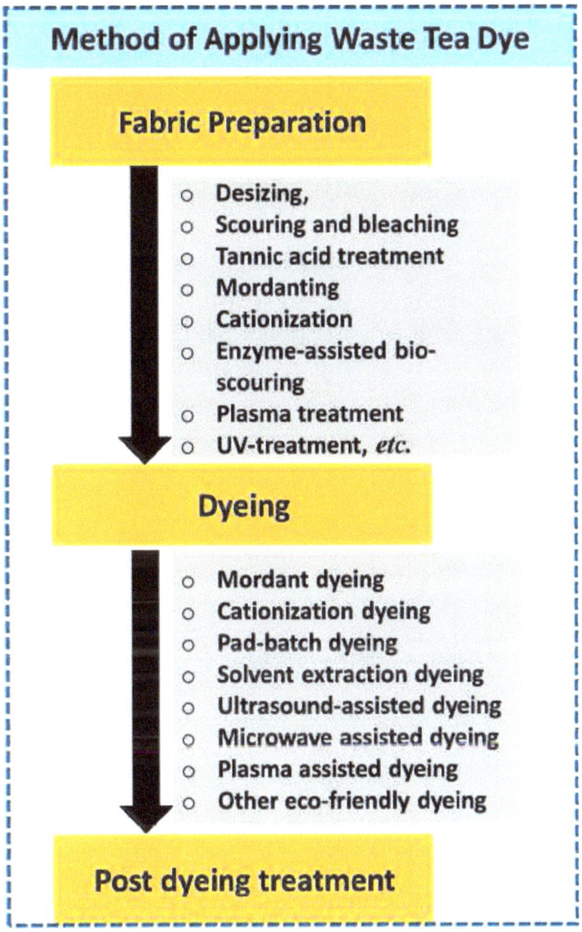

Fig. (1). Outline of the method of applying waste tea dye.

METHOD FOR APPLYING WASTE TEA DYE

Fabric Preparation

Proper fabric preparation is essential to ensure effective dye absorption. Before dyeing, impurities such as oil, dirt, and dust must be removed from the fabric surface. In woven cotton fabrics, starch is commonly applied to the warp and weft yarns during weaving to strengthen the yarns and prevent breakage. However, this starch hinders dye penetration by blocking fabric pores and causing dye aggregation. Therefore, desizing is necessary to convert insoluble starch into soluble glucose through hydrolytic or oxidative processes. This improves capillary action and ensures more uniform dye uptake.

One common method involves treating the fabric with a sodium hydroxide (NaOH) solution containing detergent and wetting agents. The fabric is immersed in this solution, which helps break down and remove impurities, improving dye penetration. After treatment, the fabric is thoroughly rinsed with distilled water to remove any residual chemicals, enhancing its dye absorbance [20].

Alternatively, fabric preparation can be carried out using a 70% ethanol solution. In this method, the fabric is immersed in the ethanol solution for half an hour, allowing it to dissolve and remove surface contaminants. After immersion, the fabric is rinsed with distilled water for five minutes and then air-dried at room temperature. For a more thorough cleaning effect, this washing process can be repeated two to three times, ensuring a cleaner fabric surface and better dye adherence [21].

After scouring and bleaching, the fabric may also be treated with tannic acid in water, particularly before mordanting. Tannic acid enhances dye uptake in cotton by introducing bonding surfaces lost during desizing and bleaching. It compensates for the removal of attaching groups on the fiber surface, thus creating new bonding edges for dye attachment. Additional eco-friendly pre-treatments can further improve dyeability. For example, enzyme-assisted bio-scouring uses pectinase or other cellulolytic enzymes to selectively remove non-cellulosic impurities from cotton and protein fibers, increasing hydrophilicity and enhancing dye absorption without harsh chemicals. Similarly, UV treatment modifies the fiber surface by introducing reactive functional groups, creating more dye absorption sites and improving dye-fiber interactions [22]. Depending on the dyeing technique, specific pretreatments are required at this stage: for cationization dyeing, the fabric must be cationized; for mordant dyeing, it must be mordanted; and for plasma-assisted dyeing, plasma treatment is applied. Detailed procedures for these methods are discussed in the respective sections below.

Dyeing Using Waste Tea Dye

Mordant Dyeing

Mordants play a crucial role in tea dyeing by facilitating the binding of dye molecules to fabric fibers, thereby improving dye uptake and color fastness. Metal mordants, the most commonly used type, form insoluble complexes with tea polyphenols and fiber functional groups such as hydroxyl and amino groups, enhancing the interaction between dye and fiber and producing varied hues depending on the metal used [18, 22, 23]. Plant-based mordants such as turmeric, pomegranate peel, and henna act similarly by introducing additional phenolic or carbonyl groups that enhance cross-linking with both the dye and the fiber, resulting in improved shade depth and fastness properties [19, 22].

When combined with factory tea waste dye, the mordant dyeing method shown in Fig. (**2a**) employs ferrous sulfate ($FeSO_4$) and copper sulfate ($CuSO_4$) as metallic mordants to improve dye uptake and fixation properties. As a result, the color fastness of the dyed fabric is improved [24 - 26]. Dyes are initially extracted from the tea waste generated by the factory, as shown in Fig. (**2b-c**). From 650 grams of factory tea waste, 94.25 grams of dyes are extracted. A single-bath procedure is used, wherein the material (jute), mordant, and dye are immersed in the same bath. The dyeing of desized jute fabric is carried out following the procedure outlined in Table **1**, with a dyeing temperature of 100°C maintained for 2 hours and a material to liquor (M:L) ratio of 1:10. Occasional shaking is done to mix dye and mordant properly and for better dye fixation on the jute fabric surface. The dyeing solution's pH is between 5.5 and 6.5 to optimize the mordant's activity. The proportions of dye and mordant added to the weight of the fabric (owf) were 16% and 10%, respectively. After dyeing, the fabric is rinsed with distilled water and dried at room temperature.

Table 1. Recipe of mordant dyeing using factory tea waste as dye.

Material	Amount (g)/ L
Extracted dye	16 grams
Distilled water	1 Liter
Desized jute fabric	100 grams
Metallic mordant	10 grams
***M: L: 1:10, *pH:5.5- 6.5**	*Dyeing Temperature: 100°

Nevertheless, pre-mordanting and post-mordanting can also be employed to dye jute fabric. The metallic mordant is employed to mordant the jute desized fabric during the pre-mordanting process. After mordanting, the fabric is subjected to an additional bath, during which the dyeing process is finalized at 100°C for 2 hours. The M:L ratio remains unaltered. The color fastness to light, cleansing, and abrasion is improved due to the enhanced dye absorption achieved through pre-mordanting. The pre-mordanting procedure is environmentally beneficial because the dye solution can be utilized multiple times. After dyeing, mordanting is performed in the post-mordanting procedure to achieve a broader spectrum of shades, after the final hue is observed. The temperature, time, pH, and M:L ratio of the dyeing process remain unaltered. The mordanting process resulted in the dyed jute fabric shown in Fig. (**2d-f**) exhibiting excellent color fastness to washing and rubbing and good light fastness properties. The fabric's color strength (K/S) value is increased due to metallic mordanting. The interaction between tea dye, mordant, and jute fabric was verified through ATR-FTIR analysis. The tea extract showed characteristic polyphenolic bands, including broad O–H stretching at 3305–3200 cm^{-1}, C–H stretching at 2921–2850 cm^{-1}, and aromatic C=C/C=O vibrations between 1739–1450 cm^{-1} [27, 28]. After dyeing, the mordanted jute fabric exhibited new or intensified peaks at 1600–1500 cm^{-1} and a distinct band around 1033 cm^{-1}, indicating the presence of bound polyphenolic structures. In addition, slight shifts in the O–H and C=O stretching regions reflected metal-mediated chelation between mordants and dye molecules. These spectral changes—absent in undyed fabric—confirm the formation of stable dye–mordant–fiber complexes, supporting enhanced dye fixation and improved color fastness [29].

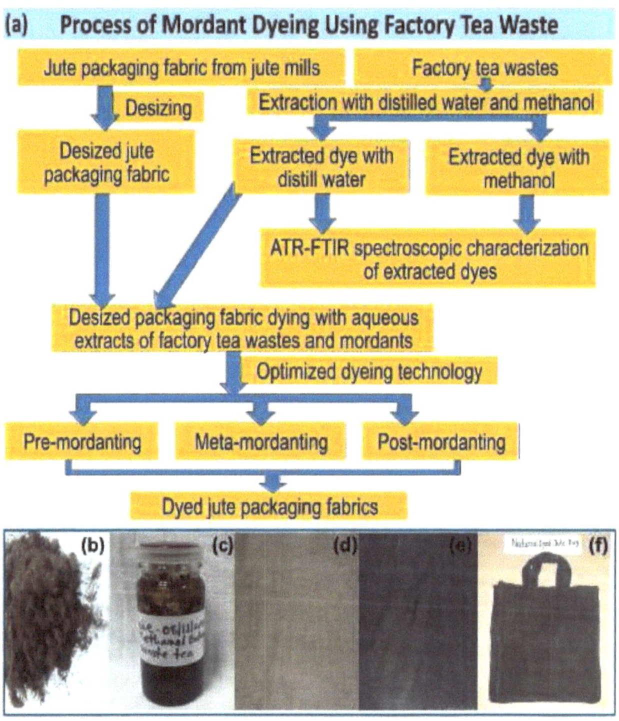

Fig. (2). (a) Process of mordant dyeing using factory tea waste, (b) Factory tea waste, (c) Extracted dye of factory tea waste, (d) Undyed jute packaging fabric, (e) Dyed jute bag [29].

The fabrics dyed with tea waste through pre, meta, and post-mordanting demonstrate remarkable color fastness characteristics, including resistance to light, washing, and rubbing. This is attributed to the extracted hydroxyl groups (OH) from the tea and the hydroxyl groups (OH) in the jute fabrics, which form coordinate bonds with the metallic mordant [30, 31]. Consequently, the color strength of the dyed jute fabrics (K/S) is improved.

Borah *et al.* demonstrated the mordant dyeing of mulberry silk utilizing dye extracted from green tea (*Camellia sinensis*) leaves. The mordant dyeing of silk demonstrated excellent dyeing properties at 80°C for 45 minutes. The color's resistance to washing, rubbing, and light exposure is improved. The fabrics showed antimicrobial reduction and UV resistance [15]. Deo *et al.* employed pre-, meta-, and post-mordanting techniques to dye cotton and jute fabrics utilizing aqueous tea extract as the dye source in a separate investigation. A commercial sample of black tea was used for the dye extraction. To extract the dye, 5 grams of commercially available tea powder were combined with 100 mL of water, agitated, and heated for half an hour. The resulting mixture was filtered and utilized as the dye. The pre-mordanting method involved immersing the fabrics, specifically cotton and jute, in an aqueous solution of metallic mordants, including alum, copper sulphate ($CuSO_4$), and ferrous sulphate ($FeSO_4$), for 45 minutes at 30°C. All mordanted fabrics were subsequently dyed at 90°C for 30 minutes utilizing a beaker dyeing machine, maintaining a material to liquor (M:L) ratio of 1:30. Following the dyeing process, the fabrics are thoroughly rinsed with water, and the washing procedure is conducted using nonionic soap at a temperature of 90°C. The dyed fabrics undergo a rinsing process with cold water before being dried at ambient temperature [32].

The meta-mordanting method involved immersing the fabrics in a bath that contained both a mordant and the dye extract, with the dyebath carefully maintained at a pH of 4. The temperature was elevated to 90°C over 30 minutes and maintained for 1 hour. The fabrics underwent a rinsing process, were treated with nonionic soap at a temperature of 60°C, followed by a thorough wash with water, then squeezed and dried. The post-mordanting method involved an initial dyeing process without using a mordant, which was subsequently followed by a separate mordanting phase in a bath containing the mordant at a controlled temperature of 30°C for 45 minutes. The dyed fabrics demonstrated superior wash fastness and good light fastness, where the wash and light fastness were carried out using ISO 2 and AATCC method 16-193 on a Sunset CPS, respectively. The fabrics' color strength (K/S) value has been enhanced [32].

Through the application of three dyeing techniques—pre-mordanting, meta-mordanting, and post-mordanting—utilizing six different mordants (alum, $FeSO_4$,

$ZnSO_4$, $CuSO_4$, $MgSO_4$, and Na_2SO_4) on wool fabric, Hasan *et al.* demonstrated that the post-mordanting method yields superior color fastness properties, where black tea waste was used to extract the dye. The raw black tea waste was collected from the cafeteria. This includes enhanced resistance to washing, rubbing, and light, and an improved color strength (K/S) value. The dyeing process is concluded at 85°C for 35 minutes, and the meta-mordanting is achieved by rinsing the fabric with nonionic detergent at 85°C. The dyeing process involved soaking 480 grams of black tea waste in 1 L of water to facilitate dye extraction, where the extraction time was varied from 2 to 8 hours. Twenty grams of dye were collected from 480 grams of raw black tea waste. The pre-mordanting and post-mordanting dyeing temperatures and times are 30°C and 45 minutes, respectively [33]. Liu *et al.* demonstrated that natural mordants, such as turmeric and tannic acid, significantly improved the color fastness of silk fabric dyed with tea extract, specifically regarding wash, rubbing, and light resistance. Dyes were extracted from black tea leaves, and the dye was diluted 5, 10, 30, and 50 times afterextraction. The ideal pH for achieving maximum color strength (K/S) is 7.87 for turmeric at a pH of 7, and 6.32 for tannic acid at a pH of 3.

Additionally, the optimal amount for achieving the highest color strength (K/S) is 0.025g/L for turmeric and 5 g/L for tannic acid. The GB/T 3921-2008 method is employed to determine the color fastness of silk fabric at a temperature of 50°C after washing. The color fastness to rubbing in silk fabric is evaluated using the GB/T3920-2008 method, which involves 10 strokes in 10 seconds and a rubbing cloth made of 100% cotton with a specific brightness. The GB/T8427-2008 method was also employed to evaluate the color fastness of cotton fabric to light. The dyed silk fabric exhibited good color fastness against wash, rubbing, and light [34].

Cationization Dyeing

The process of cationization dyeing is shown in Fig. (**3a**). A cationizer, which includes quaternary ammonium and a reactive group that interacts with cellulose, is used for pretreatment in the cationization dyeing process. This chemical process generates the cationic side of the fabric surface, which is attracted to the anionic dye by the electrostatic attraction between the fabric surface and the dye, as shown in Fig. (**3b**). The minimal rinsing and after-washing required for dyeing are made possible by the strong dye-fiber interaction created by cationic cellulose. Additionally, a fiber with a high affinity for dye may demonstrate vibrant colors and brightness following multiple washes [35].

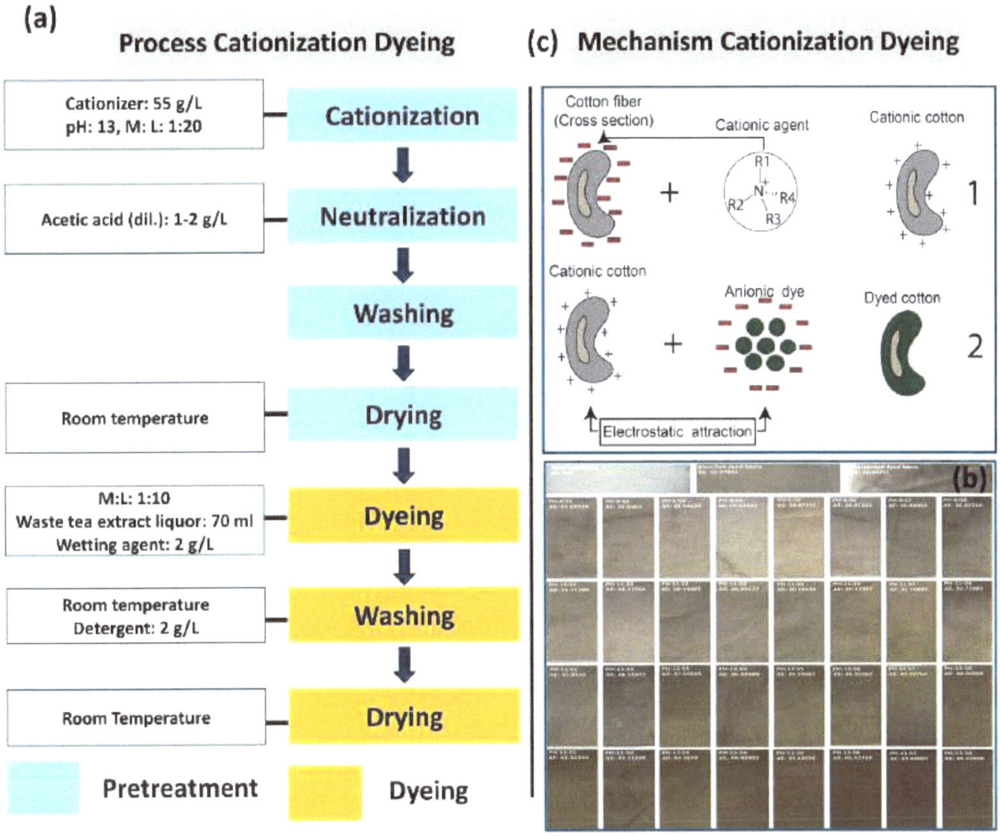

Fig. (3). (a) Process of cationization dyeing, (b) Mechanism of cationization dyeing [35], (c) Samples of bleached, dyed fabric, mordanted dyed fabric, and cationized dyed fabrics at different pH and concentrations of CHPTAC [16].

The pretreatment is initially conducted using a cationizer (CHPTAC, 3-chloro-2-hydroxypropyl trimethyl ammonium chloride) in an alkaline medium with a material-to-liquor ratio (M:L) of 1:20. A lower molecular weight facilitates the penetration of CHPTAC into fiber pores. Consequently, uniform dyeing is observed. A wetting agent is employed to enhance wettability. At the same time, acetic acid (CH_3COOH) is utilized for neutralization. The fabric is subsequently rinsed and washed with water, then dried at room temperature. Following pretreatment, the fabric is dyed using dye extracted from waste tea bags for 60 min at 90°C, with a material-to-liquid ratio of 1:10. The dye extraction process utilizes black spent tea bags collected from local tea stalls for dye extraction. In this extraction process, 5.17±0.5 grams of dye were extracted from 30 grams of black tea waste, where the extraction efficiency was 15.56-18.9%. A wetting agent is employed in the dyeing process to improve the wettability of the fabric.

The dyed fabric is subsequently removed, washed with detergent, and air-dried at room temperature. The wash fastness and rubbing fastness were evaluated using the standard methods ISO 105-C06 and ISO 105-X12:2001(E), employing a crock meter (SDL ATLAS LTD, USA) at pH levels of 12 and 13. For 72 hours, a Q-SUN XE-2 Xenon machine was used to perform the sweat and light fastness tests according to the standard method ISO 105-E04:2013 and AATCC TM16. The process demonstrates excellent color fastness against washing, rubbing, and perspiration, achieving a maximum exhaustion percentage of 88.6% at pH 13 for cationized fabrics, as shown in Fig. (**3c**), with a cationizer concentration of 55 g/L [16]. Using CHPTAC as a cationizer in this dyeing method raises significant concerns due to its toxicity. During the cationization process, it transforms into 2,3-epoxypropyl trimethylammonium chloride (EPTAC), which is recognized as a carcinogen [36].

Pad-stream Dyeing

The pad-stream dyeing method represents an innovative approach for dyeing cotton fabric, utilizing dye derived from oolong tea leaves *via in-situ* polymerization. The extraction of dye from oolong tea leaves is shown in Fig. (**4a**). The oolong tea leaves are powdered and combined with water in a 1:2 ratio. Following a 1-hour boiling period, the resultant solution is employed as a dye in the pad-stream dyeing process. Fig. (**4b**) shows that this method involves immersing the fabric in 200 mL of water for 30 seconds, then padding, steaming, and dyeing. The padding mangle functions with three dips and three nips, with a steaming duration of 5 min. Following padding and steaming, the dyed fabric undergoes washing with soap to eliminate any unfixed dyes on the surface. The fabric is subsequently dried at 70 °C [14].

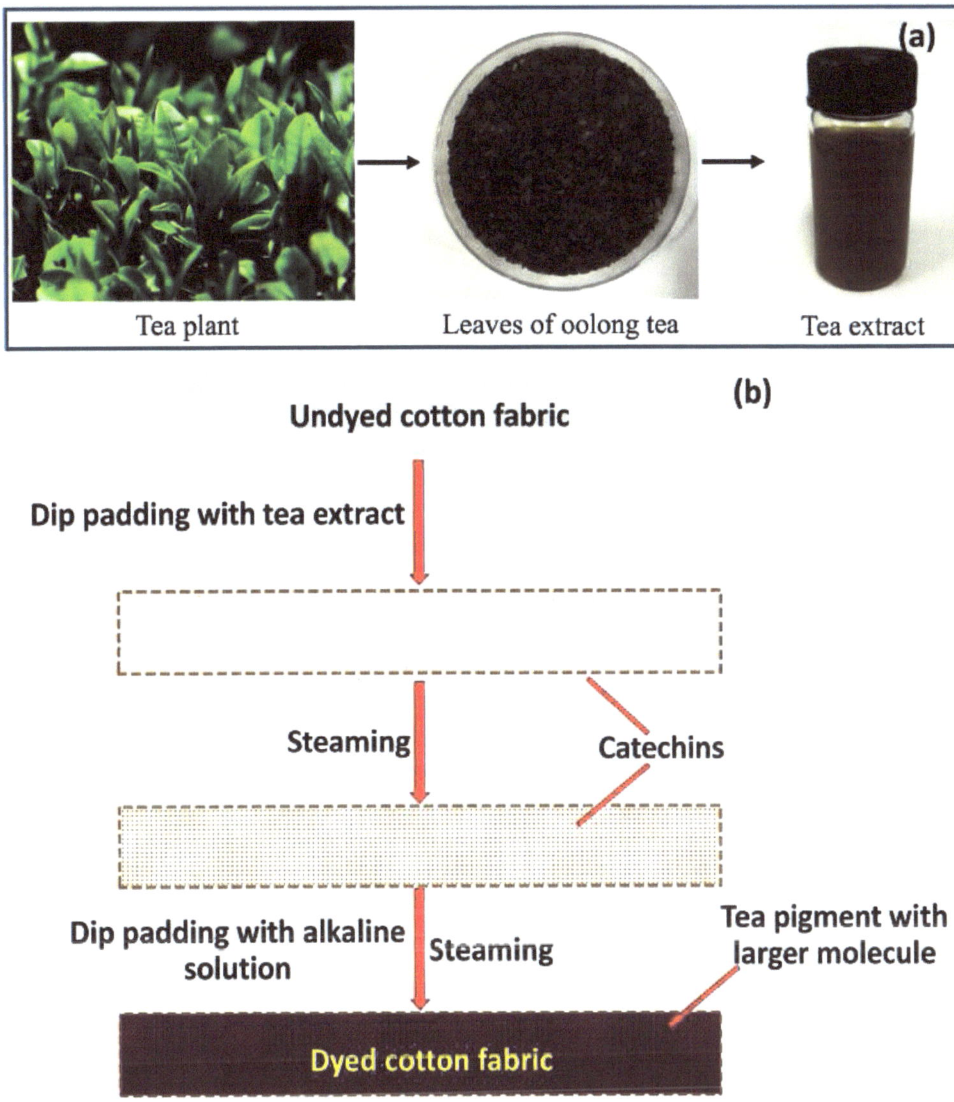

Fig. (4). (a) Extraction of tea dye from oolong tea leaves [28], (b) Process of cotton dyeing using oolong tea leaves extract *via in-situ* polymerization [14, 37].

The dyed fabric exhibited superior color fastness to washing, rubbing, light, and perspiration. The assessment of color fastness for dyed fabric subjected to washing with soap was conducted following the ISO 105-C10 standard. The washing fastness was evaluated using a 5g/L soap solution at a temperature of 40 °C for 30 minutes. The methodologies ISO 105-X12, ISO 105 E04, and ISO 105-

B02 assessed color fastness to rubbing, perspiration, and light exposure, respectively. *In-situ* polymerization improves the color strength of dyed fabric. The steaming process enhances the concentration of catechin molecules. Various oxidation processes facilitate the conversion of these compounds into catechin ortho-quinones, characterized by a complex molecular structure. The oxidation polymerization occurring within the fabric structure under alkaline conditions enhances the color fastness properties of the fabrics [14].

Solvent Extraction Dyeing

Tea stem waste collected from the tea plantation, containing phenolic compounds, is utilized for dye extraction in the simultaneous dyeing process [38]. Tea stem waste is initially combined with 200 mL of sodium carbonate (Na_2CO_3) in a conical flask and shaken for 1 hour using an oscillator. Extraction occurs under alkaline conditions. The extracted solution serves as a dye for silk fabric, with a material-to-liquor ratio (M:L) of 1:50 and a dyeing rate of 2°C/min. Following dyeing, the fabric is washed with water and dried at room temperature [38].

The dyed silk fabric undergoes post-mordanting to enhance fixation and dye uptake properties, utilizing metallic mordants such as titanium sulfate, ferrous sulfate, ferric sulfate, and potassium persulfate. The M: L ratio is 1:50, with a 2 g/L dosage rate. Mordanting occurs at 50°C for 30 min. Following mordanting, the fabric is washed with water to eliminate excess mordant from its surfaces and then dried at room temperature, as shown in Fig. (**5**). The dyed fabric demonstrates superior wash and rubbing fastness at a controlled pH of 3 and a temperature of 60°C. The dyed fabric exhibits flame retardancy due to the oxidative polymerization of polyphenols under alkaline conditions, transforming the polyphenols into macromolecular forms. The fabric possesses antimicrobial reduction properties. The dip dyeing of silk fabric demonstrates multifunctional properties. Thus, it is called simultaneous dyeing and bioactive functionalization [39].

Xu *et al.* used the solvent extraction method to dye cotton fabric using oolong tea extracts as dye. Initially, the oolong tea stems collected from tea farmers are subjected to washing and drying before being ground to minimize particle size, thereby enhancing the surface area of the dye particles. The grinding process takes one minute to convert the solid stems into a powdered state. The dye is subsequently extracted from the powder under optimal conditions, with a material to liquor ratio of 1:34, a duration of 57 minutes, and a temperature of 87°C. The concentration of polyphenolic compounds is maximized under this extraction method [40]. Before the dyeing process, the cotton undergoes scouring and bleaching to ensure complete wetting and uniform absorption capabilities, followed by pre-mordanting. The dyeing process is finalized by applying the direct dyeing technique utilizing dye extracted from oolong. The dyeing process has been meticulously optimized, with the parameters set at a temperature of 90°C, a pH of 7.9, and a duration of 80 minutes. The dyed fabric demonstrates superior color fastness when subjected to washing, exposure to light, and rubbing. The color fastness of the dyed organic cotton fabric against rubbing, washing, sweat, and light was evaluated following the relevant ISO international standards (ISO 105-X12:2001, ISO 105-C10(2):2006, ISO 105-E04:2013, and ISO 105-B01:2014). The K/S value, indicative of color strength in dyed cotton fabric, shows notable enhancement [41]. The overall extraction and dyeing process with undyed and dyed fabric is shown in Fig. (**6a-d**).

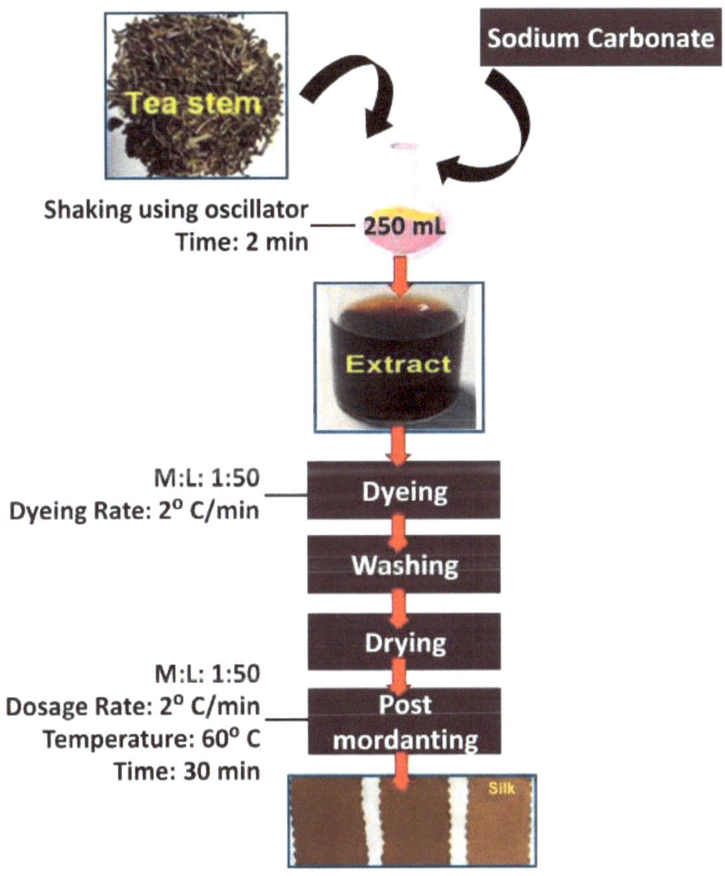

Fig. (5). Solvent extraction dyeing of silk fabric.

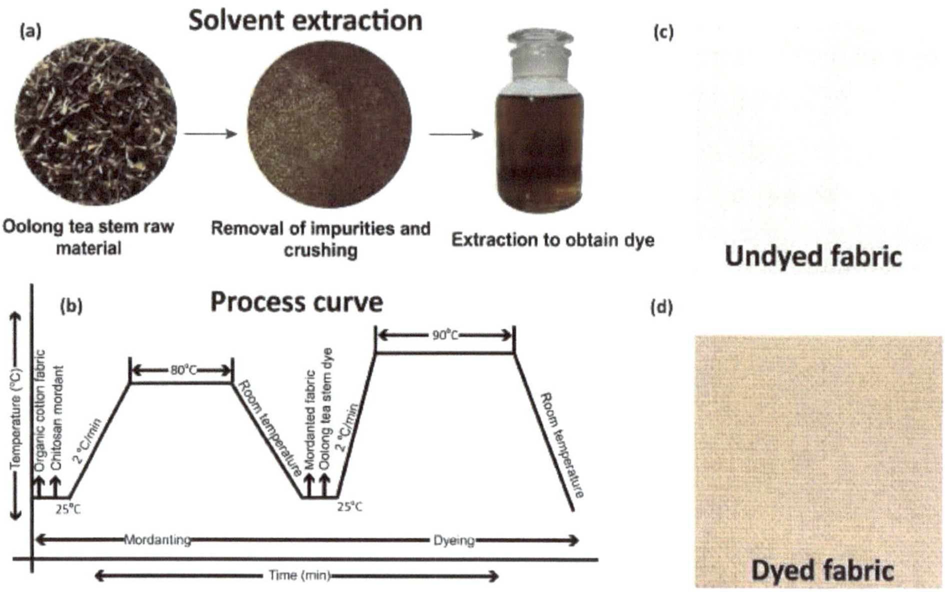

Fig. (6). (a) Dye extraction from tea stems using the solvent extraction method, (b) dyeing of cotton fabric, (c-d) undyed and dyed cotton fabric [41].

Ultrasound-assisted Dyeing

Ultrasound-assisted dyeing effectively utilizes sound waves to extract dye particles from different materials [42]. The extraction procedure is conducted at controlled temperatures using specific solvents to enhance efficiency [43]. Initially, black tea waste is washed to eliminate dust and impurities, followed by various drying methods, including oven and sunlight. The materials undergo milling to reduce their size, thereby increasing the surface area of the black tea waste. The black tea waste is soaked in distilled water, followed by ultrasound-assisted extraction in a water bath. The dyeing process involves soaking 480 grams of black tea waste in 1 Liter of water to facilitate dye extraction. Sonication of the sample was conducted at 20 – 25 kHz for 30 min at 60°C. At an extraction duration of 8 hours, utilizing 300 mL of solvent, the extraction process yielded an average extract volume of 10.67 mL. The concentration of theaflavins in the extract was determined to be 5.22 mg/L.

A fabric sample dyed using this extract exhibited a deep greenish coloration (Fig. 7), indicating the effectiveness of the extracted theaflavins in imparting color. The dyed fabric's visual assessment suggests that the extract's natural pigment successfully bonded with the textile substrate. Mordanting enhances dye uptake and fixation at 50ºC for 15-20 min. Copper sulfate serves as a mordant, and following the mordanting process, the sample is rinsed with cold water to eliminate any excess mordant. The cotton fabric is subsequently dyed using dye extracted from black tea waste. Following dyeing, the fabric undergoes washing with water and is subsequently dried at room temperature, as shown in Fig. (7) [33].

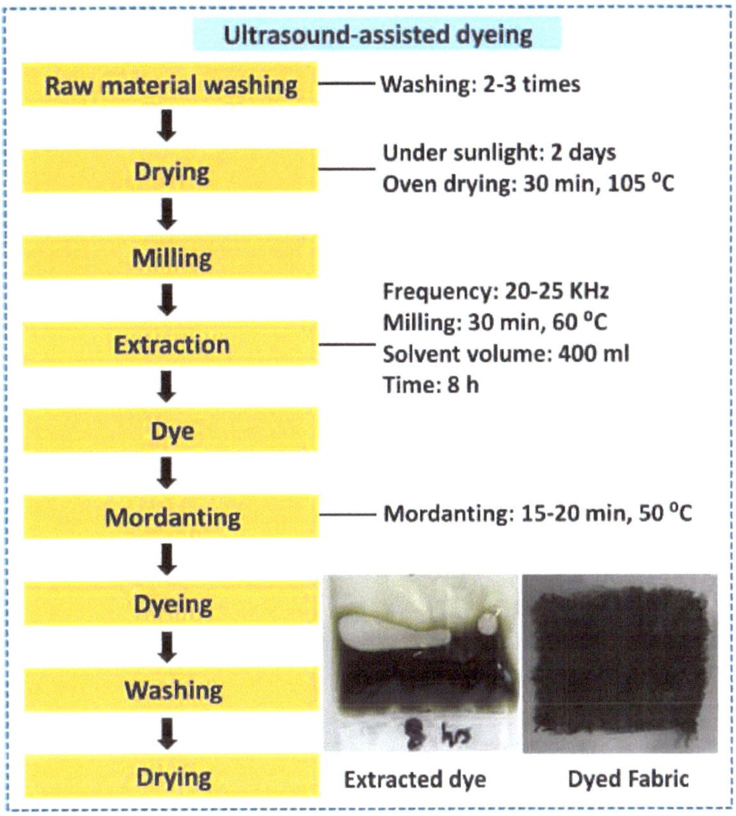

Fig. (7). Ultrasound-assisted dyeing [33].

The optimal dye extraction time is 8 hours from black tea waste, while a solvent volume of 300 mL yields the highest color intensity. The dyeing occurs at a controlled temperature of 50-60°C, utilizing 0.5% metallic mordant. The dyeing effluents exhibit reduced contamination, signifying sustainable dyeing practices [33].

Microwave-assisted Dyeing

Microwave irradiation is an economical dyeing technique that improves dye uptake, color intensity, and fixation on fabric surfaces [44, 45]. The dyed fabric exhibits excellent fastness properties. The microwave dyeing method involves extracting dye from tea leaves through microwave irradiation. Tea leaves are gathered and then boiled in an acidic medium using a material-to-liquid (M:L) ratio of 1:30. The boiling procedure lasts 30 min. The extraction is subjected to microwave irradiation for 6 min. The fabric is dyed using the extract solution at an M:L ratio of 1:35, maintained at 80°C for 50 min. The dyed fabric is mordanted to enhance the color intensity and sustainability, and aluminium sulfate (alum) and ferrous sulfate ($FeSO_4$) are used as mordants. The dyed fabric shows excellent color fastness properties, increasing the color strength (K/S) value [4]. The fabric dyed with pre- and post-mordanting techniques exhibited remarkable color fastness characteristics when ferrous sulphate served as the mordant. The presence of the metal-dye complex in the wool fabric enhances the overall color fastness properties, as shown in Fig. (**8a**) [4, 46]. Applying a bio-mordant in the microwave-assisted dyeing method yields enhanced tint strength, alongside introducing a novel shade. The interaction occurs due to the (COOH) and amide linkages present in the wool fabric, which engage with the (OH) groups found in the natural tea dye as well as the (OH) groups in curcumin molecules, as shown in Fig. (**8b**). Furthermore, their conjugation with both colorant and bio-mordant enhances the value of the colorant, ensuring it remains resistant to washing during the laundering process. Consequently, bio-mordant has outstanding colorimetric characteristics [47, 48].

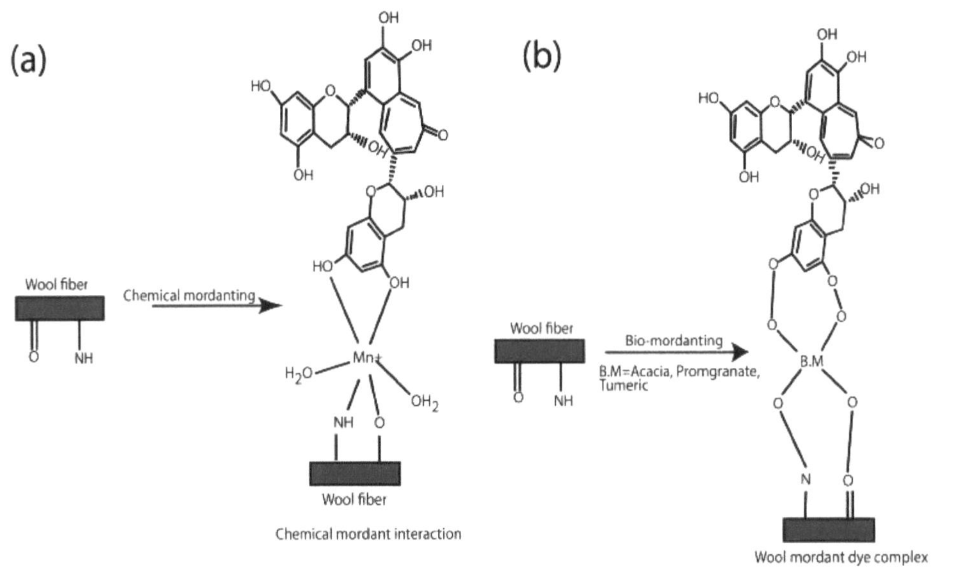

Fig. (8). (a) Chemical mordanting, (b) bio-mordanting mechanism using microwave-assisted dyeing of wool fabric using tea extract as dye [4].

Adeel *et al.* employed microwave irradiation to extract dye from black tea leaves and dyed cotton fabric using this dye. At first, the black tea waste powder was extracted in an acidic medium with microwave irradiation for 6 min. The cotton fabric was dyed for 30 min with a material-to-liquor ratio of 1:30. Following dyeing, bio-mordanting was performed to enhance the fabric's color strength and dye uptake percentage. The dyed fabric demonstrated superior color fastness and increased color strength [1].

Plasma-assisted Dyeing

The plasma-assisted dyeing process involves fabric treatment with plasma, demonstrating greater efficiency than traditional desizing methods [49]. The process further optimizes the efficiency of scouring and bleaching, thereby significantly enhancing the dyeing efficiency of the fabric. The plasma treatment enhances the capacity for water and dye absorption. The tensile properties of the fabric, which are reduced during the dyeing process, can be improved by applying short-term plasma treatment [50, 51]. The overall plasma treatment on textile substrates, including cotton and wool, is shown in Fig. (**9a**).

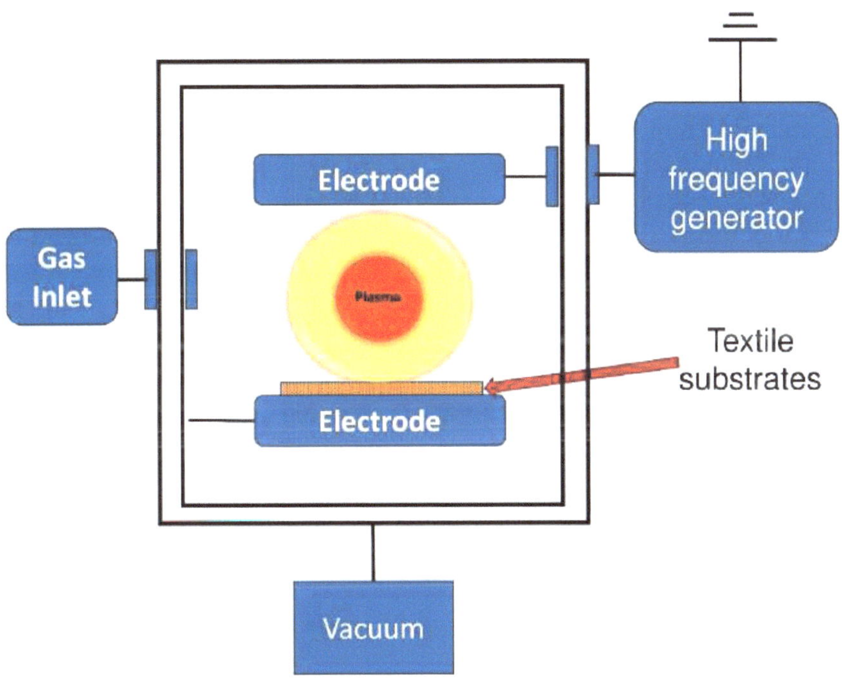

Fig. (9). Plasma treatment on textile substrates [53].

The initial step involves the surface modification of the cotton fabric utilizing a radio frequency (RF), low-pressure Inductively Coupled (IC) plasma system. The RF generator operates at the standard industrial frequency of 13.56 MHz, delivering a power output of 100W. The cotton samples underwent treatment with O_2 plasma for 10 and 30 seconds and NH_3 plasma for intervals ranging from 1 to 300 seconds. After surface modification, the dye is extracted from green tea and curcumin (Natural Yellow 3) at 95°C for 1 hour. Before extraction, the particle size of green tea and curcumin is minimized through grinding to optimise the surface area. The dyeing process of the cotton fabric was successfully concluded, with the material to liquor (M: L) ratio set at 1:50, a temperature maintained at 90°C, and a duration of 30 minutes. The color strength properties of ammonia (NH_3) plasma-treated cotton fabric are significantly improved when the treatment duration is set to 300 seconds. Applying NH_3 plasma treatment on cotton fabric significantly enhances the adsorption rate, improving colorimetric properties [52].

Other Eco-Friendly Dyeing Methods

Several innovative methods have been developed to improve the sustainability and performance of tea polyphenol dyeing, going beyond conventional techniques. One such approach is enzyme-assisted dyeing, in which laccase or crude enzymes produced by *Aspergillus niger* catalyze the oxidative polymerization of tea polyphenols into stable pigments [54]. These enzymatically formed pigments exhibit higher fixation rates on protein fabrics, such as wool and silk, particularly under acidic conditions, where theaflavins are more stable. The dyeing process involves preparing a buffer solution of tea polyphenols at the desired pH, adding the enzyme, and incubating the mixture to allow polymerization. Fabrics are then immersed in the dye solution, typically at a liquor ratio of 1:50, and dyed at elevated temperatures for 60 minutes. Enzyme-assisted in situ polymerization not only improves color intensity and fastness [55] but also enhances functional properties such as antibacterial, antioxidant, and UV protection, all without the use of toxic metal mordants [56].

Another eco-friendly approach is UV-assisted dyeing, where protein fabrics are first exposed to ultraviolet radiation to increase surface hydrophilicity and introduce reactive functional groups. These modifications improve the adsorption of tea polyphenol dyes and can be combined with meta-mordanting techniques, using mild mordants such as $CuSO_4$, to achieve deeper shades and improved color fastness [57]. The UV treatment typically lasts for 10 minutes, followed by dyeing at 95°C for 70 minutes in an acidic medium, which favors the stability of theaflavin and enhances dye-fiber interactions.

Pad–dry–cure techniques with bio-mordants also represent a cleaner alternative for tea polyphenol dyeing. In this method, cotton or other cellulose-based fabrics are impregnated with a mixture of tea polyphenols and natural catalysts such as citric acid or sodium dihydrogen phosphate. The fabric is padded to achieve a specific wet pick-up, dried at 95°C, and then cured at 140°C. This process promotes esterification between the hydroxyl groups of cellulose and the carboxyl groups of citric acid, as well as strong bonding between tea polyphenols and the fabric surface. The resulting dyed fabrics exhibit balanced improvements in color strength, wrinkle recovery, and fastness properties while maintaining the original fabric strength [58].

Additionally, residual liquor dyeing is a sustainable method that utilizes leftover tea polyphenol solutions from previous dyeing cycles. Protein and cotton fabrics can be repeatedly dyed with residual liquor, which contains sufficient pigment concentration for multiple cycles. The dyeing mechanism involves initial absorption of micromolecular tea polyphenols onto the fiber surface, followed by in situ polymerization into macromolecular pigments. The bonding occurs through hydrogen bonding, van der Waals forces, and electrostatic interactions, depending on the fiber type. Residual liquor dyeing maximizes resource utilization, reduces waste, and consistently produces fabrics with high color strength and excellent fastness properties [59].

Collectively, these eco-friendly methods demonstrate that tea polyphenols can be effectively applied in textile dyeing while reducing the environmental impact associated with conventional chemical mordants. By employing enzymatic polymerization, UV treatment, bio-mordants, and residual liquor techniques, sustainable and functional coloration of both protein and cellulosic fabrics can be achieved with improved dye-fiber interactions, color intensity, and wash durability.

Post-dyeing Treatment

A post-dyeing treatment is essential for securing dye molecules to the fabric surface. This treatment prevents color bleeding, improves the fabric's appearance, and enhances color fastness properties. Post-drying treatments include soaping, rinsing, washing, and drying. Dyed fabrics undergo washing with liquid soap or detergent (3g/L) at 90 °C for 10 min to eliminate unfixed dye particles from the fabric surfaces. The dyed fabric is dried at room temperature after washing. A drying machine is utilized for drying at elevated temperatures [14, 25, 42].

The dyeing methods utilizing tea waste utilize the properties of tea polyphenols. Nevertheless, they vary significantly in their mechanisms, scales, and associated choices. Mordanting provides enhanced wash and light fastness through metal coordination. However, it requires significant water usage and results in metal-rich effluents. Cationization results in remarkable exhaustion and brightness through the electrostatic charging of cellulose, although it depends on hazardous quaternary ammonium compounds. Pad-stream processing incorporates oxidized catechins in a low-liquor, continuous operation for swift throughput, yet it requires specialized padding-steaming equipment. The solvent extraction process produces multifunctional finishes, such as antimicrobial and flame-retardant properties, using alkaline phenol recovery and metal mordanting. However, this method is characterized by high liquor ratios and necessitates multiple stages. Ultrasound-assisted extraction facilitates the rapid release of pigments while simultaneously decreasing wastewater toxicity; however, it necessitates prolonged sonication. Microwave-assisted dyeing significantly reduces processing time and allows for creating innovative bio-mordant colors for small batches, yet it requires specialized microwave reactors. Plasma-assisted pretreatment improves fiber wettability, uptake, and strength with minimal chemical use but involves considerable capital investment. Equipment costs, effluent management, processing efficiency, and end-use performance must be considered to determine the most sustainable and effective method for dyeing using extracted dye from tea waste.

CONCLUSION

This chapter demonstrates that waste tea is a promising natural dye with strong potential for sustainable textile coloration. A comprehensive framework for its application has been presented, beginning with essential fabric preparation processes such as desizing, scouring, bleaching, tannic acid treatment, mordanting, cationization, and various bio-based or physical surface modification techniques, including enzyme, UV, and plasma treatments. These preparatory steps enhance fiber reactivity and facilitate better interaction with the polyphenolic compounds present in waste tea extracts.

Multiple dyeing approaches ranging from conventional mordant dyeing to advanced techniques such as solvent extraction dyeing, ultrasound- and microwave-assisted dyeing, pad–dry–cure processes with biomordants, plasma-assisted dyeing, and other eco-friendly alternatives were discussed to illustrate the versatility and adaptability of waste tea dyes across different fiber types. These methods improve dye uptake, shade depth, and color fastness while significantly reducing chemical use and environmental impact. Post-dyeing treatments further contribute by effectively removing unfixed dye and enhancing wash durability. Integrating such green technologies reinforces the viability of tea polyphenols as a renewable and high-performance colorant for modern textile applications. Nonetheless, issues, including the standardization of tea waste sources, optimization of extraction and application parameters, the long-term durability under diverse environmental conditions, and the ecological impact of mordants necessitate additional investigation.

ACKNOWLEDGMENTS

The authors wish to recognize the application of AI tools, like Grammarly, during the writing process to enhance the clarity and language of the manuscript. The authors thoroughly reviewed and modified the content as necessary and accept full responsibility for the content of the chapters. Additionally, they want to emphasize that AI was not utilized in their chapters to create or modify images, aid in data analysis, or contribute to their research.

REFERENCES

[1] Adeel S, Ahmad M, FazalurRehman, *et al.* Eco-friendly dyeing of cotton fabric with waste tea leaves-based tannin natural dye. Glob NEST J 2021; 23(3): 365-9. [http://dx.doi.org/10.30955/gnj.003456]

[2] Yusuf M, Shabbir M, Mohammad F. Natural Colorants: Historical, Processing and Sustainable Prospects. Nat Prod Bioprospect 2017; 7(1): 123-45. [http://dx.doi.org/10.1007/s13659-017-0119-9] [PMID: 28093670]

[3] Yadav S, Tiwari KS, Gupta C, Tiwari MK, Khan A, Sonkar SP. A brief review on natural dyes, pigments: Recent advances and future perspectives. Results in Chemistry 2023; 5: 100733. [http://dx.doi.org/10.1016/j.rechem.2022.100733]

[4] Adeel S, Azeem M, Habib N, *et al.* Sustainable Application of Microwave Assisted Extracted Tea Based Tannin Natural Dye for Chemical and Bio-Mordanted Wool Fabric. J Nat Fibers 2023; 20(1): 2136322. [http://dx.doi.org/10.1080/15440478.2022.2136322]

[5] Mittal J. Permissible Synthetic Food Dyes in India. Resonance 2020; 25(4): 567-77. [http://dx.doi.org/10.1007/s12045-020-0970-6]

[6] Ashfaq M, Ali A, Nawaz Tahir M, *et al.* Synthetic approach to achieve halo imine units: Solid-state assembly, DFT based electronic and non linear optical behavior. Chem Phys Lett 2022; 803: 139843. [http://dx.doi.org/10.1016/j.cplett.2022.139843]

[7] Munawar KS, Ali S, Muhammad S, *et al.* Synthesis, crystal structure, Hirshfeld surface analysis, DNA binding, optical and nonlinear optical properties of Schiff bases derived from o-aminophenol. J Mol Struct 2023; 1274: 134427. [http://dx.doi.org/10.1016/j.molstruc.2022.134427]

[8] Kargar H, Fallah-Mehrjardi M, Behjatmanesh-Ardakani R, *et al.* Spectroscopic investigation, molecular structure, catalytic activity with computational studies of a novel Pd(II) complex incorporating unsymmetrical tetradentate Schiff base ligand. Inorg Chem Commun 2022; 142: 109697. [http://dx.doi.org/10.1016/j.inoche.2022.109697]

[9] Rabbi MBU, Haque S, Bedoura S. Advancements in Synthesis, Immobilization, Characterization, and Multifaceted Applications of Silver Nanoparticles: A Comprehensive Review. Heliyon 2024; 10(24): e40931. [http://dx.doi.org/10.1016/j.heliyon.2024.e40931]

[10] Alegbe EO, Uthman TO. A review of history, properties, classification, applications and challenges of natural and synthetic dyes. Heliyon 2024; 10(13): e33646. [http://dx.doi.org/10.1016/j.heliyon.2024.e33646] [PMID: 39040261]

[11] Siddiqui MW, Sharangi AB, Singh JP, *et al.* Antimicrobial Properties of Teas and Their Extracts *in vitro*. Crit Rev Food Sci Nutr 2016; 56: 1428–39. [http://dx.doi.org/10.1080/10408398.2013.769932]

[12] Sukemi , Pratumyot K, Srisuwannaket C, Niamnont N, Mingvanish W. Dyeing of cotton with the natural dye extracted from waste leaves of green tea (*Camellia sinensis* var. *assamica*). Color Technol 2019; 135(2): 121-6.
[http://dx.doi.org/10.1111/cote.12381]

[13] Che J, Yang X. A recent (2009–2021) perspective on sustainable color and textile coloration using natural plant resources. Heliyon 2022; 8(10): e10979.
[http://dx.doi.org/10.1016/j.heliyon.2022.e10979] [PMID: 36281409]

[14] Ren Y, Fu R, Fang K, *et al.* Dyeing cotton with tea extract based on *in-situ* polymerization: An innovative mechanism of coloring cellulose fibers by industrial crop pigments. Ind Crops Prod 2019; 142: 111863.
[http://dx.doi.org/10.1016/j.indcrop.2019.111863]

[15] Borah S, Bhuyan PM, Sarma B, Hazarika S, Gogoi A, Gogoi P. Sustainable dyeing of mulberry silk fabric using extracts of green tea (Camellia sinensis): Extraction, mordanting, dyed silk fabric properties and silk-dye interaction mechanism. Ind Crops Prod 2023; 205: 117517.
[http://dx.doi.org/10.1016/j.indcrop.2023.117517]

[16] Hossain T, Das SC, Akhtarujjaman M, Uddin MA, Bedoura S. Eco-friendly dyeing of cotton fabric using used tea bag extracts: A comprehensive study on pH, fabric cationisation, and computational analysis. Current Research in Green and Sustainable Chemistry 2023; 7(August): 100387.
[http://dx.doi.org/10.1016/j.crgsc.2023.100387]

[17] Triwiswara M, Indrayani L. Utilization of black tea waste as natural batik dyes on cotton and silk. IOP Conf Ser Earth Environ Sci 2020; 456(1): 012051.
[http://dx.doi.org/10.1088/1755-1315/456/1/012051]

[18] Wang P, Wu H, Zheng X, *et al.* High-binding-fastness dye from functional extracts of Keemun black tea waste for dyeing flax fabric. Color Technol 2022; 138(3): 255-65.
[http://dx.doi.org/10.1111/cote.12587]

[19] Hayat T, Adeel S, Fazal-ur-Rehman , *et al.* Waste black tea leaves (*Camelia sinensis*) as a sustainable source of tannin natural colorant for bio-treated silk dyeing. Environ Sci Pollut Res Int 2022; 29(16): 24035-48.
[http://dx.doi.org/10.1007/s11356-021-17341-5] [PMID: 34822080]

[20] Hauser P. Fabric Finishing: Pretreatment/Textile Wet Processing. In: Sinclair R, Ed. Textiles and Fashion. Woodhead Publishing 2015; pp. 459-73.
[http://dx.doi.org/10.1016/B978-1-84569-931-4.00018-0]

[21] 2025.
[http://dx.doi.org/10.1021/acsomega.4c07348]

[22] Xu H, Song H. A Systematic Review of Tea Dyeing in the Textile Industry: Research Trends, Dyeing Performance, Functional Characteristics, and Application Potential. AATCC Journal of Research 2025; 12(1): 24723444251314916.
[http://dx.doi.org/10.1177/24723444251314916]

[23] Moiz A, Aleem Ahmed M, Kausar N, Ahmed K, Sohail M. Study the effect of metal ion on wool fabric dyeing with tea as natural dye. J Saudi Chem Soc 2010; 14(1): 69-76.
[http://dx.doi.org/10.1016/j.jscs.2009.12.011]

[24] Jabar JM, Adedayo TE, Odusote YA. Green, eco-friendly and sustainable alternative in dyeing cotton fabric using aqueous extract Mucuna slonaei F dye: effects of metal salts pre-mordanting on color strength and fastness properties. Current Research in Green and Sustainable Chemistry 2021; 4: 100151.
[http://dx.doi.org/10.1016/j.crgsc.2021.100151]

[25] Uddin MG. Effects of Different Mordants on Silk Fabric Dyed with Onion Outer Skin Extracts. Journal of Textiles 2014; 2014(1): 1-8.

[http://dx.doi.org/10.1155/2014/405626]

[26] Ding Y, Freeman HS. Mordant dye application on cotton: optimisation and combination with natural dyes. Color Technol 2017; 133(5): 369-75. [http://dx.doi.org/10.1111/cote.12288]

[27] Ebrahimian A, Saberikhah E. Biosorption of methylene blue onto Foumanat tea waste: equilibrium and thermodynamic studies. Cellul Chem Technol 2013; 47: 657-66.

[28] Ren Y, Gong J, Wang F, *et al.* Effect of dye bath pH on dyeing and functional properties of wool fabric dyed with tea extract. Dyes Pigments 2016; 134: 334-41. [http://dx.doi.org/10.1016/j.dyepig.2016.07.032]

[29] Sarker MKU, Haque MM, Hasan MR, Sultana S, Ray SK, Shaikh MAA. Utilization of factory tea (*Camellia sinensis*) wastes in eco-friendly dyeing of jute packaging fabrics. Heliyon 2024; 10(10): e30948. [http://dx.doi.org/10.1016/j.heliyon.2024.e30948] [PMID: 38779026]

[30] Janani L, Hillary L, Phillips K. Mordanting Methods for Dyeing Cotton Fabrics with Dye from Albizia coriaria Plant Species. Int J Sci Res Pub 2014; 4(10). Available from: www.ijsrp.org

[31] Sultana S, Akter K, Sarker MKU, Bhuiyan RH, Haque MM, Islam MR. Color Fastness and Tensile Properties of Cotton Fabric Dyed with Extract from Albizia Procera Sawdust. Fibers Polym 2022; 23(10): 2820-7. [http://dx.doi.org/10.1007/s12221-022-4707-x]

[32] Deo HT, Desai BK. Dyeing of cotton and jute with tea as a natural dye. Color Technol 1999; 115(7-8): 224-7. [http://dx.doi.org/10.1111/j.1478-4408.1999.tb00360.x]

[33] Yanfei R, Gong J, Wang F, *et al.* Effect of dye bath pH on dyeing and functional properties of wool fabric dyed with tea extract. Dyes and Pigments 2016; 134: 334-41. [http://dx.doi.org/10.1016/J.DYEPIG.2016.07.032]

[34] Huang X, Liu C. Analysis of Natural Mordant to B. mori Silk Fabrics Dyeing with Tea Extract IOP Conf Ser Mater Sci Eng vol. 585: 012044.2019; [http://dx.doi.org/10.1088/1757-899X/585/1/012044]

[35] Correia J, Rainert KT, Oliveira FR, de Cássia Siqueira Curto Valle R, Valle JAB. Cationization of cotton fiber: an integrated view of cationic agents, processes variables, properties, market and future prospects. Cellulose 2020; 27(15): 8527-50. [http://dx.doi.org/10.1007/s10570-020-03361-w]

[36] Choudhury AKR. Coloration of Cationized Cellulosic Fibers–A Review. AATCC Journal of Research 2014; 1(3): 11-9. [http://dx.doi.org/10.14504/ajr.1.3.2]

[37] Sang S, Lambert JD, Ho CT, Yang CS. The chemistry and biotransformation of tea constituents. Pharmacol Res 2011; 64(2): 87-99. [http://dx.doi.org/10.1016/j.phrs.2011.02.007] [PMID: 21371557]

[38] Stepanova AV, Kotina EL, Tilney PM, Van Wyk BE. Leaf and stem anatomy of honeybush tea (Cyclopia species, Fabaceae). S Afr J Bot 2012; 82: 123-8. [http://dx.doi.org/10.1016/j.sajb.2012.07.001]

[39] Cheng TH, Liu ZJ, Yang JY, Huang YZ, Tang RC, Qiao YF. Extraction of Functional Dyes from Tea Stem Waste in Alkaline Medium and Their Application for Simultaneous Coloration and Flame Retardant and Bioactive Functionalization of Silk. ACS Sustain Chem& Eng 2019; 7(22): 18405-13. [http://dx.doi.org/10.1021/acssuschemeng.9b04094]

[40] Yang L, Dai X, Meng Y, Zhou L, Yue X. Comparative Study on Dyed Fabrics of Tea and Tea Stem Based on Kansei Engineering. Sustainability (Basel) 2023; 15(2): 1134. [http://dx.doi.org/10.3390/su15021134]

[41] Xu H, Yang C, Song H. Eco-Friendly Dyeing and Functional Finishing of Organic Cotton Using Optimized Oolong Tea Stems (Agricultural Waste) Through Response Surface Methodology. Molecules 2025; 30(3): 509.
[http://dx.doi.org/10.3390/molecules30030509] [PMID: 39942614]

[42] Banna BU, Mia R, Hasan MM, Ahmed B, Hasan Shibly MA. Ultrasonic-assisted sustainable extraction and dyeing of organic cotton fabric using natural dyes from *Dillenia indica* leaf. Heliyon 2023; 9(8): e18702.
[http://dx.doi.org/10.1016/j.heliyon.2023.e18702] [PMID: 37560636]

[43] Zumahi SMAA, Arobi N, Taha H, *et al.* Extraction, optical properties, and aging studies of natural pigments of various flower plants. Heliyon 2020; 6(9): e05104.
[http://dx.doi.org/10.1016/j.heliyon.2020.e05104] [PMID: 33024874]

[44] Arain RA, Ahmad F, khatri Z, Peerzada MH. Microwave assisted henna organic dyeing of polyester fabric: a green, economical and energy proficient substitute. Nat Prod Res 2021; 35(2): 327-30.
[http://dx.doi.org/10.1080/14786419.2019.1619721] [PMID: 31140300]

[45] Ticha MB, Slama N, Dhouibi N, Boudokhane C, Dhaouadi H. Bark Residues Recovery of Juglans Regia. L For the Dyeing of Wool Fabrics : Development of Microwave-assisted Extraction and Dyeing Processes. J Nat Fibers 2022; 19(13): 5282-96.
[http://dx.doi.org/10.1080/15440478.2021.1875370]

[46] Ayele M, Tesfaye T, Alemu D, Limeneh M, Sithole B. Natural dyeing of cotton fabric with extracts from mango tree: A step towards sustainable dyeing. Sustain Chem Pharm 2020; 17: 100293.
[http://dx.doi.org/10.1016/j.scp.2020.100293]

[47] Gong K, Rather LJ, Zhou Q, Wang W, Li Q. Natural dyeing of merino wool fibers with *Cinnamomum camphora* leaves extract with mordants of biological origin: a greener approach of textile coloration. J Textil Inst 2020; 111(7): 1038-46.
[http://dx.doi.org/10.1080/00405000.2019.1698228]

[48] Zhang W, Wang X, Weng J, *et al.* Eco-dyeing and functional finishing of wool fabric based on *Portulaca oleracea* L. as colorant and *Musa basjoo* as natural mordant. Arab J Chem 2022; 15(2): 103624.
[http://dx.doi.org/10.1016/j.arabjc.2021.103624]

[49] Zaisheng Cai , Yiping Qiu , Chuyang Zhang , Hwang YJ, Mccord M. Effect of Atmospheric Plasma Treatment on Desizing of PVA on Cotton. Text Res J 2003; 73(8): 670-4.
[http://dx.doi.org/10.1177/004051750307300803]

[50] Haque ANMA, Smriti SA, Farzana N, Siddiqa F, Islam MA. Fuzzy Modelling for Prediction of Bursting Strength of Knitted Cotton Fabric using Bleaching Process Variables. AATCC Journal of Research 2019; 6(1): 29-37.
[http://dx.doi.org/10.14504/ajr.6.1.5]

[51] Peng S, Gao Z, Sun J, Yao L, Qiu Y. Influence of argon/oxygen atmospheric dielectric barrier discharge treatment on desizing and scouring of poly (vinyl alcohol) on cotton fabrics. Appl Surf Sci 2009; 255(23): 9458-62.
[http://dx.doi.org/10.1016/j.apsusc.2009.07.058]

[52] Gorjanc M, Mozetič M, Vesel A, Zaplotnik R. Natural dyeing and UV protection of plasma treated cotton. Eur Phys J D 2018; 72(3): 41.
[http://dx.doi.org/10.1140/epjd/e2017-80680-9]

[53] M. Naebe, A. N. M. A. Haque, and A. Haji, “The Effect of Plasma Treatment on Dyeing of Natural Fibers,” in Innovative and Emerging Technologies for Textile Dyeing and Finishing, 2021, pp. 191–212.
[http://dx.doi.org/10.1002/9781119710288.ch7]

[54] Wang F, Gong J, Zhang X, Ren Y, Zhang J. Preparation of Biocolorant and Eco-Dyeing Derived from

Polyphenols Based on Laccase-Catalyzed Oxidative Polymerization. Polymers (Basel) 2018; 10(2): 196.
[http://dx.doi.org/10.3390/polym10020196] [PMID: 30966232]

[55] Gong J, Wang F, Ren Y, Li Z, Zhang J, Li Q. Preparation of biomass pigments and dyeing based on bioconversion. J Clean Prod 2018; 182: 301-12.
[http://dx.doi.org/10.1016/j.jclepro.2018.01.219]

[56] Garg H, Singhal N, Singh A, Khan MD, Sheikh J. Laccase-assisted colouration of wool fabric using green tea extract for imparting antioxidant, antibacterial, and UV protection activities. Environ Sci Pollut Res Int 2023; 30(35): 84386-96.
[http://dx.doi.org/10.1007/s11356-023-28287-1] [PMID: 37365356]

[57] Chen C, Li M, Wang C, Fu S, Yan W, Chen C. Meta-mordant Dyeing with *Camellia sinensis* (L.) O. Ktze var. waldensae (S.Y.Hu) Chang (Yellow-bud Tea) Extract for Wool Fabrics Treated by UV Radiation. Fibers Polym 2018; 19(6): 1255-65.
[http://dx.doi.org/10.1007/s12221-018-7769-z]

[58] Maulik SR, Das D, Bhattacharya SC. Concurrent dyeing and finishing of cotton with natural colour and citric acid in the presence of NaH_2PO_4 as catalyst under thermal treatment. J Textil Inst 2011; 102(6): 491-9.
[http://dx.doi.org/10.1080/00405000.2010.491984]

[59] Wang F, Gong J, Ren Y, Zhang J. Eco-dyeing with biocolourant based on natural compounds. R Soc Open Sci 2018; 5(1): 171134.
[http://dx.doi.org/10.1098/rsos.171134] [PMID: 29410827]

CHAPTER 7

A Success Story of Transforming Ready to Drink Tea Waste into Fabric Dye

Vinitha Moolchand Thadhani[1,3,*], **Gehan Amaratunga**[2], **Syed G. Musharraf**[3] and **Naheed Akhtar**[3]

[1] *Ceylon Ayurveda Pvt. Ltd, Havelock Road, Colombo, Sri Lanka*

[2] *Department of Engineering, Electrical Engineering Division, University of Cambridge, Cambridge, UK*

[3] *H.E.J. Research Institute of Chemistry, International Center for Chemical and Biological Sciences, University of Karachi, Karachi, Pakistan*

Abstract: Numerous studies on the natural dyeing methods of tea (*Camellia sinensis*) for different textile materials can be found in the literature. Furthermore, there are only a few studies in the literature that specifically focus on the textile dyes generated from tea waste. However, for industrial applications, specialist laboratory-scale research might not be enough. This study establishes the discarded phenolic fraction of the Ready To Drink (RTD) tea manufacturing process as an effective fabric dye, at an industrial scale for cotton, linen, nylon, wool, and their blends, with an effective concentration of tea dye ranging from 1-10%, depending on the fabric type and desired colour. The method was exaggerated to obtain different shades with the required fastness properties.

Unlike acid, disperse, and reactive dyes, utilizing waste as an effective fabric dye results in negative CO_2 emissions.

Further, the main challenges of commercial application of Tea as a natural dye, namely, supply chain, cost, reproducibility, and consistency, have been overcome by utilizing waste generated from the world's largest RTD manufacturing plant. The technology has been commercialized and adopted by numerous reputable textile brands worldwide.

Keywords: Cost-effective, Chitosan, Fastness-properties, Ready to Drink, Sustainable, Tea waste.

[*] **Corresponding author Vinitha Moolchand Thadhani:** H.E.J. Research Institute of Chemistry, International Center for Chemical and Biological Sciences, University of Karachi, Karachi, Pakistan; E-mail: vinimool@yahoo.com

INTRODUCTION

The gigantic global textile industry is the second-highest greenhouse gas emitter after the oil industry. The textile industry generates around $ 1 trillion, comprises 7% of the total world exports, and employs around 35 million workers worldwide [1, 2]. Simultaneously, it is responsible for 8% of global carbon emissions and 20% of global clean water pollution, with an anticipated 50% increase in greenhouse gas emissions by 2030 [3]. Synthetic dyes are a significant contributor to the carbon footprint and water pollution associated with the textile industry. Annual synthetic dye production worldwide is estimated to be 7×10^5 tons [4], with a market share of USD 17.6 billion in 2023. It is projected to grow from USD 19.5 billion in 2024 to $21.79 billion in 2025, at a Compound Annual Growth Rate (CAGR) of 11.5% [5], despite its significant environmental impact.

Therefore, a ban on textiles based on non-eco-friendly synthetic dyes is being enforced. This has driven global research towards replacing synthetic dyes with natural colours which can come from biodegradable natural sources including from Catechu, Gamboge tree, Indigofera leaves, Tea leaves, Coffee beans, Madder root, Chestnut hulls, Walnut hulls, Mangosteen peel, Pomegranate rind, Berries, Cherries, Red & Pink Roses, Beets, Cochineal insects, Murex snails, Octopus/Cuttlefish and others [6]. However, to exploit these on a commercial level, the supply chain, cost, and, most importantly, reproducibility, to ensure consistency in production, are the main challenges.

There is ample literature reporting the use of Tea (*Camellia sinensis*) as a good source of colorant due to the presence of colored polyphenolic compounds [7-15]. Even after ample literature reports on this subject, there are very few examples of industrial applications of tea dye, mainly due to its poor wash and Light fastness properties; thus, it has to be used in conjunction with various mordants for its commercial applications [10, 16, 17].

This study describes a method for extracting an antibacterial tea dye powder from waste, specifically tea cream generated during the iced tea powder manufacturing process, as an effective fabric dye with desired fastness properties. By utilizing waste generated from Ready-to-Drink tea plants as an effective fabric dye, a negative CO_2 emission is achieved. Negative emission means reducing the amount of carbon by capturing it, in other words, extracting it from waste and storing it as a dye in textiles.

Due to its numerous health benefits, the Ready-To-Drink (RTD) tea market is expected to grow from its current CAGR of 5.4% to a CAGR of 6.12% in the coming five years [18]. During the instant tea powder manufacturing process, a significant quantity (approximately 25%) of the hot water tea extract, obtained through a hot water counter-current extraction process, is discarded as haze or tea cream. This is to ensure that RTD tea powder is soluble in cold water and suitable for mixing with ice [19]. In other words, nearly ¼ of the tea extract rich in coloured phenolic constituents is discarded to make RTD tea extract soluble in cold water. Particularly, water-soluble coloured flavonoids in tea extracts are separated from coloured glycosylated phenolic compounds during the RTD ice tea manufacturing process. Herein, we establish the fabric dyeing properties of this discarded fraction, namely tea cream, and elucidate how the discarded tannins and other glycosylated phenolic compounds can be used as an effective fabric dye with the desired fastness properties [20, 21].

THE PROCESS OF MANUFACTURING READY TO DRINK TEA (RTD)

What is Ready to Drink Tea

Ready-To-Drink Tea (RTD) powder is a concentrated tea extract that has been processed to dissolve easily in water. RTD tea requires no brewing or steeping, and thus offers greater convenience compared to loose-leaf Tea or tea bags, making it a quick and easy option for those with busy lifestyles or limited time.

Tea leaves are extracted with hot water, and the resulting hot water extract is concentrated and spray-dried to obtain tea extract in powder form, which is referred to as RTD tea powder.

Manufacturing Process of Ready-to-Drink Instant Tea

Raw Material

The main raw material used to produce RTD tea in Sri Lanka is black tea (*Camellia sinensis*) rejects, namely Broken Mixed Fannings (BMF), which are small bits of tea that are left over after higher forms of tea leaves are gathered and processed. Approximately 30,000 MT of BMF is generated for the 350,000 MT of high-quality black tea exported each year from Sri Lanka. Sri Lanka has become the primary sourcing unit for the global instant iced tea market, accounting for nearly 70% of the global supply of the Ready to Drink volumes sourced from Pepsi Lipton International, utilizing only 10,000 MT of BMF, while a further 20,000 MT of BMF remains unutilized.

Extraction

The active substances of the tea leaves are extracted using aqueous extraction in hot water in continuous extraction systems. Counter-current extraction is typically used when the solid material and the liquid (solvent) flow in opposite directions, or counter-currently. In this process, tea leaves are fed into the lower end of the extractor and are conveyed upwards by the rotation of the screw, while the extracting water flows in the opposite direction.

The BMF is extracted with hot water at 80 - 95 °C for 20 -45 min., with Liquid to Material Ration (L/M) varying from 10:1 to 30:1. The solution coming from the extractor usually contains 3-5% solids, which is measured as Brix value.

Decanting

The extracted slurry is subjected to passing through continuous decanters and clarifiers to remove the non-soluble suspended matter from the extract. In decanters, centrifugal force is used to separate large volumes of insoluble tea leaf components from tea extracts efficiently in a manner that is easy to control.

After squeezing out the solid waste from the decanter, the tea extract (A) accounts for 25-30% w of the original BMF. The tea extract (A) thus obtained is called hot water soluble Ready to Drink Tea and is used as an ingredient in tea vending machines such as Nestea. The tea extract (A) is further heated for 20-30 min at 80-95 °C in the presence of O_2, cooled to 40-45 °C, and treated with 2 g/L of hydrolyzing enzymes, such as Viscozyme (Sigma Aldrich), a cellulolytic enzyme mixture.

The resulting mixture (B), with a solid content of around 3-6%, is centrifuged. During Centrifugation, fine particles are sedimented (D) from the extracted tea liquor (Supernatant C) (Fig. **1**).

The centrifuge utilizes the sedimentation principle, where centripetal acceleration causes denser substances and particles to move outward in a radial direction. At the same time, less dense objects are displaced and move to the centre.

In the centrifugal separator, effectiveness is increased by stacking conical disks to create a large equivalent clarification arca, and the force of gravity is replaced with a centrifugal force that can be more than 5,000 times stronger than the force of gravity. Therefore, the separation will be significantly faster.

Concentration

The supernatant (C) obtained from the centrifugation process is too dilute to be passed directly to a spray dryer. Therefore, the liquor must be concentrated, typically to achieve a solid content of 20-40%. The concentration of tea extract is achieved through evaporation under reduced pressure. At high temperatures, the theaflavins in the solution are converted to thearubigins, and carbohydrates caramelize.

Spray Drying

In a spray dryer, concentrated Supernatant (C) is fed into the drying tower by a high-pressure homogenizing pump and then sprayed as a fine droplet dispersion into a stream of heated air. Air and solids move in parallel or counterflow directions, resulting in rapid drying and minimizing the damage of exposure to heat for a longer time

The dry product is collected from the bottom of the spray dryer, which is the cold water-soluble iced tea powder, used as the main ingredient to form numerous flavored ready-to-drink tea beverages.

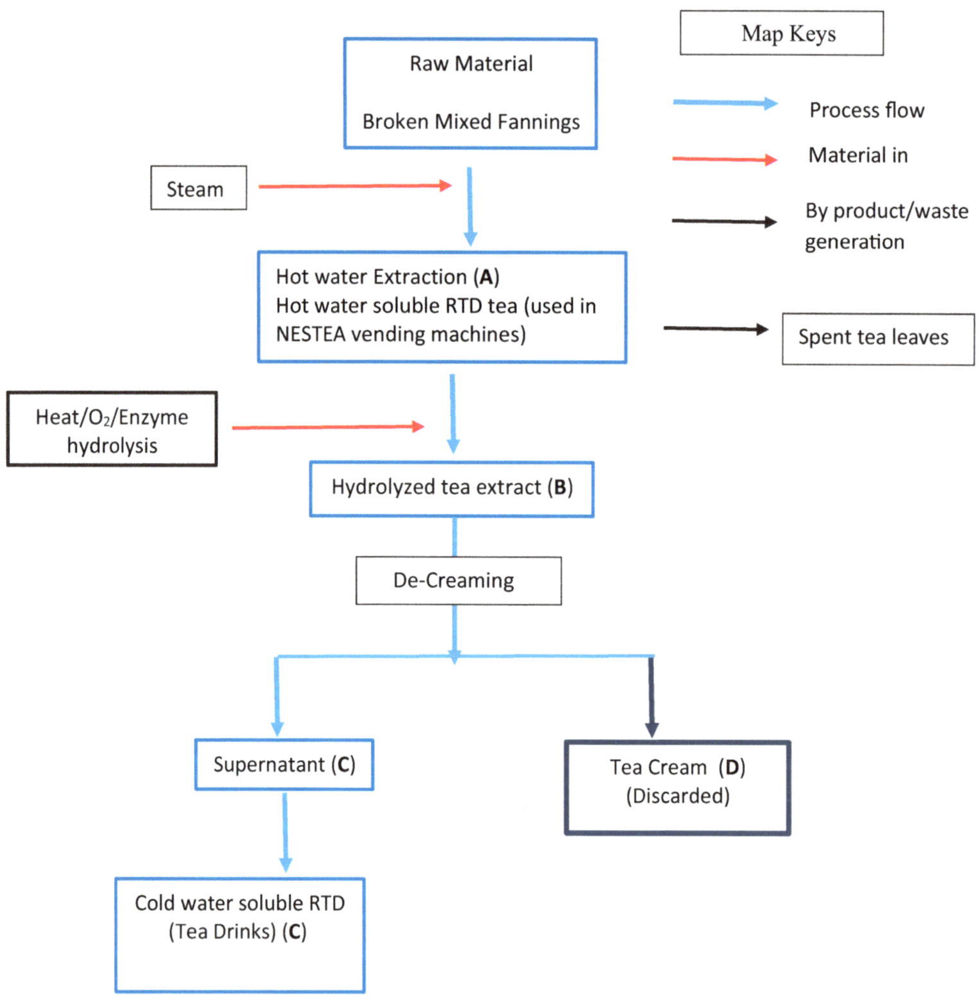

Fig. (1). Process map for the manufacturing of ready to drink tea.

PROCESS OF CONVERTING "TEA CREAM," A WASTE PRODUCT OF THE RTD TEA MANUFACTURING PROCESS, INTO A SUSTAINABLE FABRIC DYE

During the process of converting hot water soluble ready-to-drink tea extract (A), to cold water soluble ice tea powder (C), a significant quantity (amounting to about 25%) of the hot water tea extract A (Fig. **1**), is discarded as haze or tea cream (Sediment D). This is to ensure that RTD tea powder (C) is soluble in cold water and suitable for mixing with ice (10). In other words, nearly ¼ of the tea extract rich in coloured phenolic constituents, including tannins, is discarded to convert hot water soluble RTD tea extract (A) into cold water soluble RTD ice tea powder (C). This process is termed as De-Creaming.

De-Creaming

The de-creaming process in instant tea manufacturing is a crucial step for improving the quality and solubility of the cold water-soluble iced tea product. It involves removing a fraction of the tea extract, known as the "cream," which contains components like tannins that can bind to proteins, causing cloudiness or a bitter taste in the final product.

On the other hand, tannins, naturally occurring polyphenolic compounds, can be used as both mordants and dyes for fabrics. They act as a "bridge" between the dye and the fabric, enhancing color and wash fastness, especially for cellulose fibers like cotton. Tannins themselves can also impart earthy colors, ranging from beiges to browns, and can be used to deepen other natural dye colors [22].

Herein, we establish the fabric dyeing properties of this discarded fraction, namely tea cream rich in tannins and other glycosylated phenolic compounds, as an effective fabric dye [20-22].

Process of Extraction of Tea Dye from Tea Cream

The haze or tea cream (D), is filtered through a mesh of 200-400 μM, and sediment (E) with complexed glycosides with particle size above 400 μM are removed, while the filterate (F) mainly consisting of soluble phenolic compounds including tannins with solid content of 10-14%, is further evaporated to regain the solid content up to 15-20%, which is spray dried at 240 °C to obtain the powder, which is used as the tea dye [20, 21] (Fig. **2**).

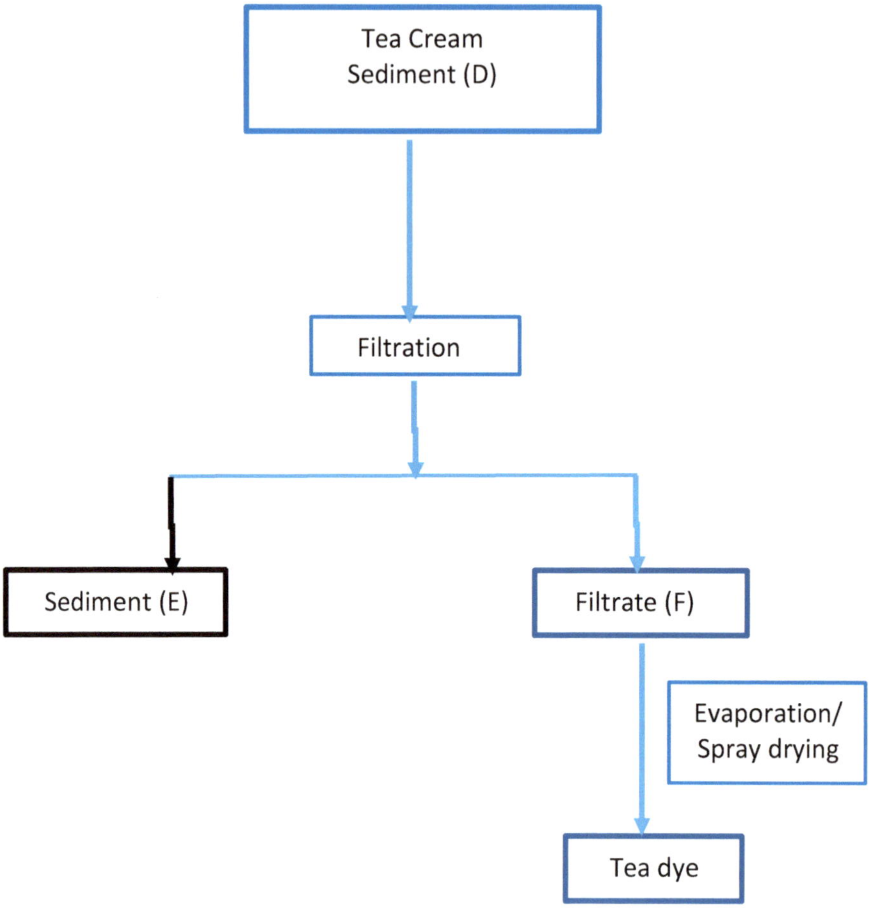

Fig. (2). Process map for the manufacturing of tea dye through waste generated during the RTD manufacturing process.

The difference in the chemical composition of (D) with respect to (B) and (C) is illustrated in the FTIR spectra (Fig. **3**).

The wavelengths numbers of FTIR spectra at 819, 1,038, 1,144, 1237, 1364, 1448, 1,516 1,604, and 1695 cm^{-1} were assigned to -C-H alkenes, -C-O alcohols, -OH aromatic, -C-N, -C=O (symmetric stretching), -C-H alkanes (bending), -C=C aromatic ring, -C=C alkenes, and -C=O (asymmetric stretching). The wave absorption band of (D) was specifically different from (B) and (C) between 1237-1364 cm^{-1} regions [20].

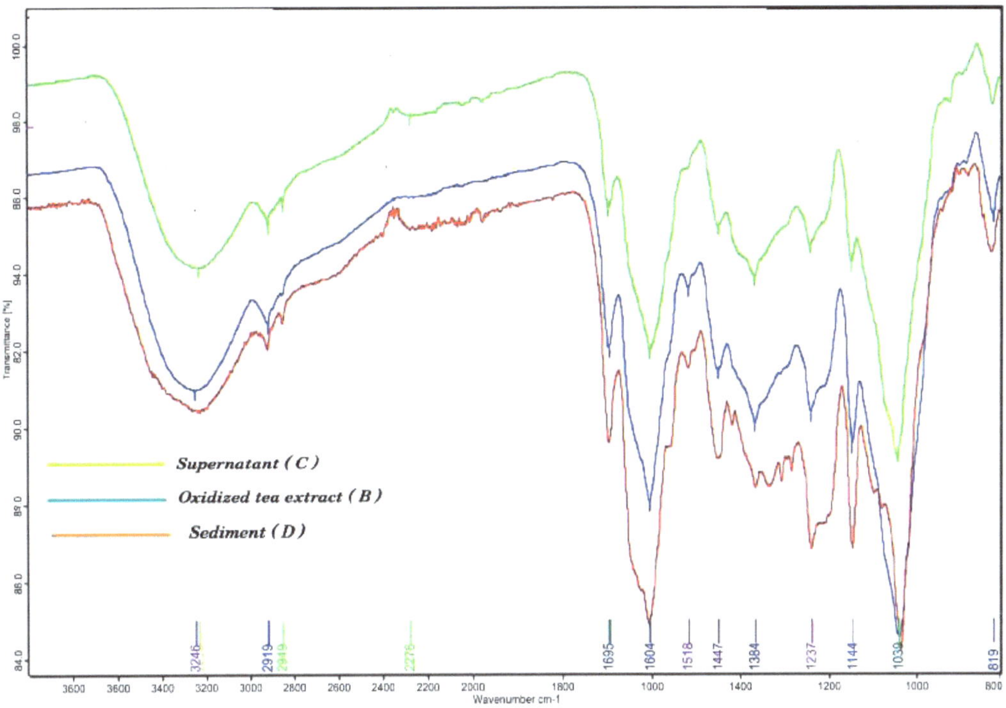

Fig. (3). FTIR spectra of different fractions of tea extracts.

The composition of tea dye (F) was analyzed using High Resolution Liquid Chromatography coupled with Quadrapole Time of Flight Mass Spectrometry (LC-QTOF-MS/MS) (Bruker Daltonic Maxis II). Compounds were investigated based on retention time, exact mass, fragmentation pattern, and quasi-molecular ion under positive mode. Composition analyses were carried out using Data Analysis 4.4 and Target Analysis 1.3 software, along with standards. MS/MS spectra were searched against the NIST MS/MS, Mass Bank of Europe, and Mass Bank of North America libraries.

CHEMICAL COMPOSITION ANALYSIS OF THE TEA DYE FORMED THROUGH THE WASTE OF THE RTD TEA MANUFACTURING PROCESS

Chemical Profiling Using LC-ESI-QTOF-MS/MS

Profiling of tea extracts A, B, C, and F was done using a Thermo Fisher UltiMate 3000 series (USA) UHPLC system, which features an autosampler, thermostat column compartment, degasser, and a binary pump, coupled to a Bruker Daltonic

MaXis II ESI-QTOF-MS instrument (Bremen, Germany). The separation was performed at a 40 °C thermostat temperature using a reverse-phase Macherey-Nagel Nucleodur C-18 gravity column with dimensions of 3.0 mm × 100 mm and a particle size of 1.8 μm. The mobile phase consisted of solvent A (deionized water + 0.1% Formic acid) and solvent B (Methanol + 0.1% Formic acid), which were sonicated for 15 minutes. The mobile phase gradient was programmed as follows: 10% B was maintained for 0-1 minute, 70% B for 2.5-4 minutes, then increased to 90% B for 6.5-8 minutes, and finally decreased again to 10% B for 9-10 minutes. Analysis of the samples was performed using an injection volume of 5 μL and a constant flow rate of 0.7 mL/min. The total run time of the analysis was 10 minutes, and a post-run of 1 minute was provided before and after gradient elution to ensure equilibration. MS data was recorded at the mass range of 100-1000 *m/z*. Sodium formate was used as a calibrant, which was maintained at a constant flow rate of 3 μL/min throughout the sample analysis [23].

For the metabolite identification of *C. sinensis,* some compounds were identified by screening MS/MS spectra in available MS/MS libraries, including NIST MS/MS (Version 2.2), the Mass Bank of Europe, and the Mass Bank of North America. While already reported compounds were also searched in the High-Resolution (HR) mass spectra of analyzed samples using Bruker Compass Target Analysis (Version 1.3) software. All the compounds were matched based on their exact masses, isotopic patterns, and MS/MS fragmentation patterns. The exact mass was matched within the 5 ppm error limit, and the isotopic pattern within the 50 m Sigma value limit. Synonyms of the compounds were searched through the Human Metabolites Database (HMDB) and PubChem.

Profiling of Samples and Identification of Compounds

Composition analysis of waste tea dye using high-resolution (LC-ESI-QTOF-MS) identified over 60 different phenolic compounds, which provide the fabric-dyeing properties for the tea cream. Further, it was also established that tea cream was rich in bulkier glycosylated flavonoid content as compared to the initial hot water extract of BMF.

Table **1** compares the intensities, which are proportional to the concentration of these compounds in the original hot water extract of Broken Mixed Fannings (BMF) (the starting material) and the tea cream extract, which is an effective natural tea dye with the required wash and light fastness properties [20, 21, 23].

Table 1. Comparison of compound intensities as for LC-QToF data, found in BMF and Tea dye.

S No.	Compound name	BMF	Spray-dried filtrate F (Tea dye)
1	Apigenin-8-C-glucoside /Vitexin	12724	20191
2	Genistein 8-C-glucoside	12724	20191
3	Isovitexin	12724	20191
4	Apigenin-8-C-glucoside-2'-rhamnoside/ Vitexin-2-rhamnoside	0	7807
5	Apigenin-7-O-glucoside	12724	20191
6	Apigenin-6-C-glucoside-7-O-glucoside/ Saponarin/ Isovitexin-7-O-glucoside	12087	17978
7	Vicinin 2	12087	17978
8	Isovitexin 2"-O-Rhamnoside	0	7807
9	Pelargonin	12087	17978
10	Neoeriocitrin	0	7807
11	5,7-dihydroxy-3-(4-hydroxyphenyl)-6,8-bis[3,4,5-trihydroxy-6-(hydroxymethyl)oxan-2-yl] chromen-4-one	6778	12234
12	8-[4,5-dihydroxy-6-(hydroxymethyl)-3-[3,4,5-trihydroxy-6-(hydroxymethyl)oxan-2-yl] oxyoxan-2-yl]-5,7-dihydroxy-2-(4-hydroxyphenyl)chromen-4-one	6778	12234
13	5,7-dihydroxy-2-(4-hydroxyphenyl)-6-[3,4,5-trihydroxy-6-(hydroxymethyl)oxan-2-yl] chromen-4-one	12724	20191
14	3-[(2S,3R,4S,5R,6S)-3,4-dihydroxy-6-methyl-5-[(2S,3R,4S,5S,6R)-3,4,5-trihydroxy-6-(hydroxymethyl)oxan-2-yl]oxyoxan-2-yl]oxy-2-(3,4-dihydroxyphenyl)-5-hydroxy-7-[3,4,5-trihydroxy-6-(hydroxymethyl) oxan-2-yl]oxychromen-4-one	0	7759
15	trans-Resveratrol	0	10480
16	CIANIDANOL	0	24565
17	Theobromine	39151	7638
18	Theophyllin	39151	7638
19	Quercetin-3-D-galactoside/ Quercetin-3-O-hexoside/ Hyperoside	20337	0
20	Quercetin-3-O-glucopyranoside/ Isoquercetin/ Hirsutrin	20337	0
21	Quercetin-4'-O-glucoside/ Spiraeoside	20337	0
22	Quercetin-7-O-rhamnoside/ Vincetoxicoside B	24042	14081
23	Quercetin-3-O-alpha-L-rhamnopyranoside	24042	14081
24	Quercitrin/ Quercetin 3-O-alpha-rhamnopyranoside	24042	14081
25	Quercetin-3-O-beta-glucopyranosyl-7-O-alpha-rhamnopyranoside	25582	0
26	Kaempferol-7-O-neohesperidoside	25207	0

(Table 1) cont.....

S No.	Compound name	BMF	Spray-dried filtrate F (Tea dye)
27	Kaempferol-3-O-rutinoside (Nicotiflorin)	25207	0
28	Kaempferol-3-O-glucoside (Astragalin)	24042	14081
29	kaempferol-7-O-hexoside/ Kaempferol-7-O-glucoside	24042	14081
30	Luteolin-4'-O-glucoside	24042	14081
31	Datiscetin-3-O-rutinoside/ Datiscin	25207	0
32	(2R,3R)-2-(3,4-dihydroxyphenyl)-3,5,7-trihydroxy-6-[(3R,4R,5S,6R)-3,4,5-trihydroxy-6-(hydroxymethyl)oxan-2-yl]-2,3-dihydrochromen-4-one	24042	14081
33	5,7-dihydroxy-2-(4-hydroxyphenyl)-3-[(2S,3R,4S,5S,6R)-3,4,5-trihydroxy-6-(hydroxymethyl)oxan-2-yl]oxychromen-4-one	24042	14081
34	Cyanidin-3-O-glucoside/ Kuromamine	24042	14081
35	Ideain/ Cyanidin-3-O-galactoside	24042	14081
36	Luteolin-7-O-glucoside/ Cynaroside	24042	14081
37	Maritimein/ Maritimetin-6-O-glucoside	24042	14081
38	Plantaginin/ Scutellarein-7-O-glucoside	24042	14081
39	Rutin/ Rutoside / Quercetin-3-O-rutinoside	25582	0
40	Delphinidin-3-O-beta-glucopyranoside	20337	0
41	Myricitrin/ Myricetin 3-O-rhamnoside	20337	0
42	Hesperidin	25582	0
43	2-(3,4-dihydroxyphenyl)-5,7-dihydroxy-3-[(2S,3R,4S,5R,6R)-3,4,5-trihydroxy-6-(hydroxymethyl)oxan-2-yl]oxychromen-4-one	20337	0
44	3,5,7-trihydroxy-2-[3-hydroxy-4-[(2S,3R,4S,5S,6R)-3,4,5-trihydroxy-6-(hydroxymethyl) oxan-2-yl]oxyphenyl]chromen-4-one	20337	0

As illustrated in Table **1**, the LC-QTOF analysis of tea dye obtained after spray drying of filterate (F) showed a higher percentage of flavones/isoflavones glycosides including but not limited to, Apigenin glycosides, Vitexin, Isovitexin, Genistein 8-C-glucoside/ Vicinin 2, Saponarin, Pelagornin, Vitexin-2-rhamnoside, Isovitexin 2”-O-Rhamnoside, 5,7-dihydroxy-3-(4-hydroxyphenyl)-6,8-bis [3,4,5-trihydroxy-6-(hydroxymethyl)oxan-2-yl] chromen-4-one, 5,7-dihydroxy-2-(4-hydroxyphenyl)-6-[3,4,5-trihydroxy-6-(hydroxymethyl)oxan-2-yl] chromen---one, 8-[4,5-dihydroxy-6-(hydroxymethyl)-3-[3,4,5-trihydroxy-6-(hydroxymethyl)oxan-2-yl] oxyoxan-2-yl]-5,7-dihydroxy-2-(4-hydroxyphenyl)chromen-4-one, 3-

[(2S,3R,4S,5R,6S)-3,4-dihydroxy-6-methyl-5-[(2S,3R,4S,5S,6R)-3,4,5-trihydroxy -6-(hydroxymethyl)oxan-2-yl]oxyoxan-2-yl]oxy-2-(3,4-dihydroxyphenyl)-5-hydroxy-7-[3,4,5-trihydroxy-6-(hydroxymethyl) oxan-2- yl]oxychromen-4-one, Neoeriocitrin, whereas lower in concatenated polymers, such as-theobromine and theophylline and also comparatively lower in flavonol glycosides namely but not limited to Quercetin glycosides, Isoquercetin glycosides, Kaempferol glycosides, Hyperoside, Nicotiflorin, Hirsutrin, Rutin, Myricitrin, Hespiridin, Astragalin, Spiraeoside, Datistin, Maritimein, Plantaginin, 2-(3,4-dihydroxyphenyl)-- ,7-dihydroxy-3-[(2S,3R,4S,5R,6R)-3,4,5-trihydroxy-6-(hydroxymethyl)oxan-2-yl]oxychromen-4-one, 3,5,7- trihydroxy-2-[3-hydroxy-4-[(2S,3R, 4S,5S,6R)-3,4,5-trihydroxy-6-(hydroxymethyl) oxan-2-yl]oxyphenyl]chromen-4-one [20, 21].

The structures of flavone glycosides and flavanol glycosides are shown in Figs. (**4** and **5**).

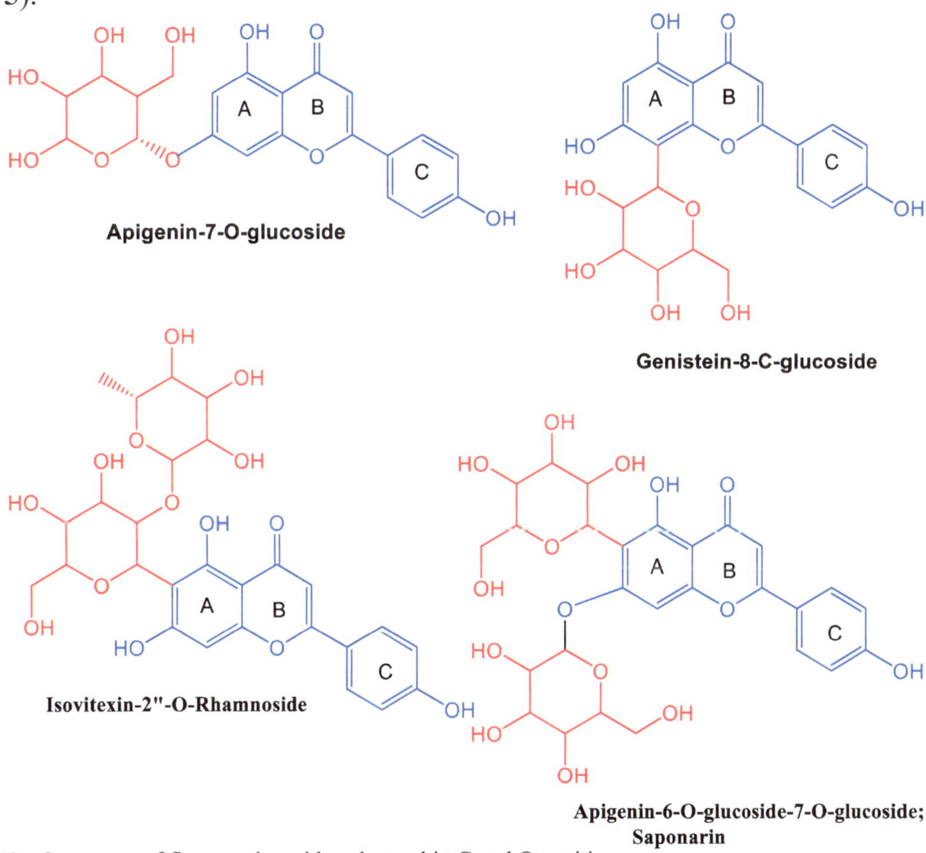

Fig. (4). Structures of flavone glycosides glycated in C and O positions.

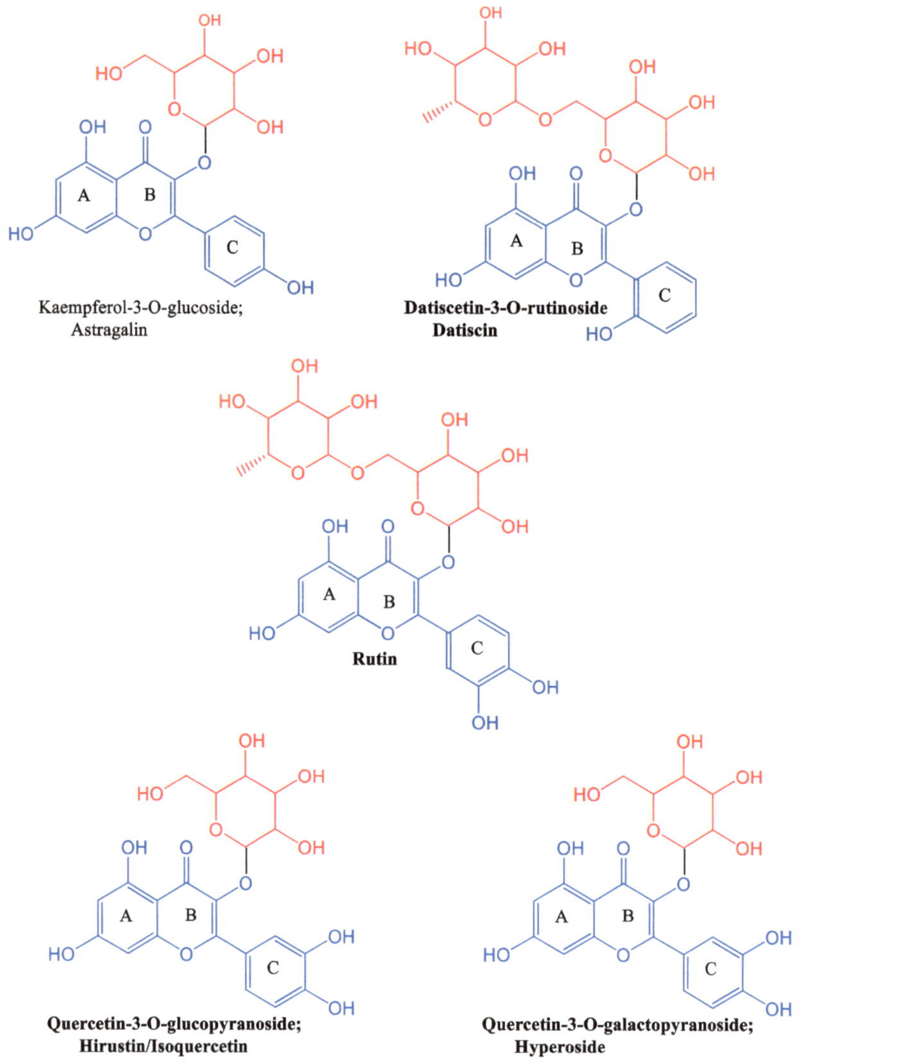

Fig. (5). Structures of flavanol glycosides.

As for Chemical Profiling Using LC-ESI-QTOF-MS/MS, the trend of distribution of identified compounds in all tea samples, namely A, B, C, and F, is shown in Fig. (**6**), which indicates that the tea dye is rich in bulkier glycosylated flavonoid content compared to the initial tea extract.

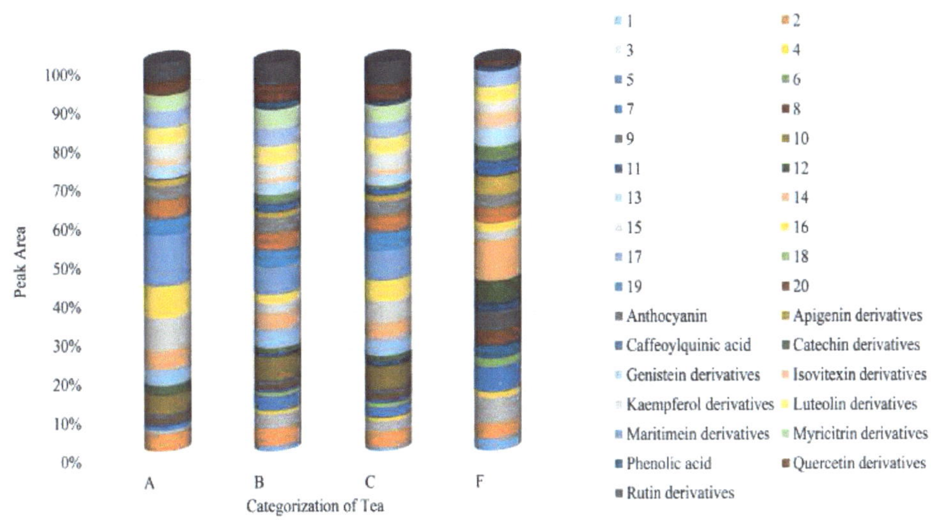

Fig. (6). Finger Print Patterns of Identified compounds in negative mode.

Hierarchical Cluster Analysis (HCA)

HCA was employed for the discrimination of A, B, C, and F tea extracts and to compare with the original hot water extract of BMF. HCA categorized the samples into classes that are closely related to each other. To visualize the distribution of different metabolites in A, B, C, and F tea extracts, Heatmap clusters were generated based on the averages of the areas of identified compounds, as log(2x), using Perseus Software (version 1.6.2.1).

Fig. (7) is the heat map, analysis of LC-MS metabolomics. The row displays the compounds, and the column represents the tea samples, namely Broken Mixed Fannings (BMF) and the tea dye. The intensity of color related to each compound showed the proportion of each metabolite in the respective sample. Compounds which significantly decreased (low intensity) are displayed in green, while compounds which significantly increased (high intensity) are displayed in red, whereas grey color shows the absence of the compound. The brightness of each color corresponded to the magnitude of the difference when compared with the average value.

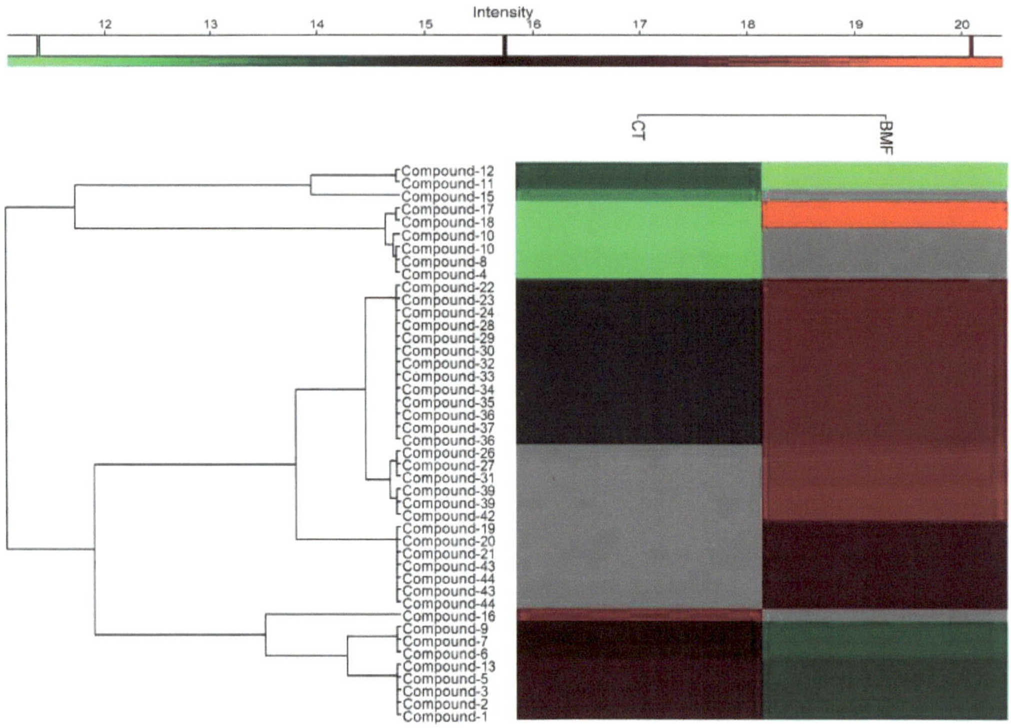

Fig. (7). Heat Map of Tea Dye as compared to the original hot water extract of BMF in -ve mode of LC-M--Q-Tof Analysis.

METHOD OF APPLICATION OF TEA DYE FORMED THROUGH THE WASTE OF THE RTD TEA MANUFACTURING PROCESS

The application of tea dye on cotton, linen, nylon, and wool was carried out using the exhaustion method. Herein, the tea dye refers to the spray-dried powder of F, as illustrated in the above process map in Fig. (**2**).

Exhaustion Method of Dyeing Cotton, Nylon, and Wool Fabrics Using Waste Tea Dye

All textile dyeing experiments were conducted using the exhaustion method in a water bath containing the tea dye extract (F). The disclosed natural dyeing process includes pretreating the fabrics with a polyamine ligand solution, such as chitosan (0.5-1.0%) [20 - 22], at a temperature between 60° and 80 °C for 15-30 minutes at pH 4. The pretreated fabric are then treated with the natural dye solution, concentration ranging from 2-10% as for the fabric used (cotton, linen, nylon wool), for sufficient time (40-45 min.) at the desired temperature (70- 80 °C for cotton and linen, and 90-95 °C for nylon and wool) to produce the required color. Unlike prior art techniques of natural dyeing, it is often necessary to choose conditions that severely damage the fiber to obtain a suitably dyed fabric. The dyeing conditions of this invention, based on RTD tea waste, are compatible with the synthetic dyeing procedures, with no additional drastic conditions or increasing the dyeing temperature or time.

Dye concentrations were determined colorimetrically by measuring at the maximum absorbance of the polyphenol dyes (λmax = 260 nm – 400 nm). A calibration curve was plotted between absorbance and concentration of the dye solution to obtain the absorbance-concentration profile for the dye. The amount of dye uptake per gram of fabric (mg/g) at any time (C_f) is calculated using a mass–balance relationship as follows.

$$C_f = \frac{(C_o - C_t)\, V}{\mathrm{W}}$$

Where C_0 is the initial dye concentration (mg/l), and C_t is the dye concentration after dyeing time t (mg/l), V is the volume of dye solution (ml), and W is the weight of fabric (g) used [14].

The adsorption capacity was determined at different operation parameters that affect the adsorption of natural polyphenol dye onto fabric. *e.g.*, initial dye concentration (5-25 g/100 g of fabric), pH (3-8), temperature (60-95 °C), ligand concentration (0.5-2 g/100g of fabric).

Generally, the dye diffusion coefficient was highest for wool and nylon and lowest for cotton. The acidic media exhibited maximum percent absorption for tea dye. The results obtained from different experiments led to the optimization of a standard recipe for a particular tea dye-mordant-fibre combination.

The optimum concentration of tea dye for dyeing cotton was found to be 5-10 g per 100 g of cotton with ligand concentration of 1g/100 g of cotton at pH 4 and temperature of 70 °C, and a dyeing time of 40 minutes, showing required light and wash fastness properties without deteriorating the quality of fabric. Whereas, for nylon, the optimum dye concentration was 1-3 g/100 g of fabric, with a ligand concentration of 0.5 g/100 g of fabric, at pH 4, and a dyeing temperature of 80-95°C for 40 min., showing good dye absorption with the required fastness properties. MLR for cotton fabric was kept at 1:20, and for nylon at 1:15. The high electron kinetic surface potential of cotton fibre when compared to nylon fibre requires larger amount of ligand as when compared to nylon fibre.

However, dyeing polyester fabric with tea dye using the exhaustion method did not yield the required fastness properties. The poor penetration of the larger molecular size tea dye components could be the reason for poor dyeability.

Use of Polyamines as Natural Mordants

The issue of using tea extracts as a biomass pigment on a commercial scale is due to the limitation that the tea compounds, which give the colour of the tea dye, such as flavan-3-ols and their concatenated polymers (such as tannins), contain many phenolic groups and are hydrophilic. Thus, these negatively charged dye molecules are poorly adsorbed to the hydrophobic fabric and tend to come into the aqueous phase. The issue is worse in the case of cotton, which has negative charge OH groups on the surface. Therefore, there is repulsion between the similarly charged polyphenolic dye of the Tea, which results in poor absorption of the dye to the fabric and does not meet the required wash and light fastness properties of the dye.

In detail, Cotton can be described chemically as 1,4-β-D-anhydroglucopyranose. Natural cellulose fibre carries a small negative charge (ζ plateau = - 1.1 mV) due to the presence of some carboxylic acid groups from oxidation at the primary hydroxylic sites. These cellulose fibers, when immersed in water, produce a negative electrokinetic potential. The negative charge on the fiber repels the anionic dye ions, consequently limiting the exhaustion of the dye bath.

Chitosan, a natural polycationic compound, can be easily adsorbed onto cellulose fibre due to its structural resemblance. The NH_2 group of chitosan can be protonated under acidic conditions to form NH_3^+, which, when applied to the fibre, reduces the negative charge carried by the surface of the fibres and ultimately reduces the electrostatic repulsion between anionic dyes, thereby improving the dye uptake. The NH_2 group of chitosan is protonated at pH 4, and a strong network of H-bonding forms between the NH_3^+ groups of chitosan and the OH groups of cellulose, reducing the negative surface charge of cellulose. Similarly, the NH_3^+ groups of chitosan bind to the carboxylic acid end of the nylon fibre. When the chitosan-treated fabric is treated with tea dye, essentially polyphenols that have been oxidized to quinone types, a Schiff base is formed between the NH_2 groups of chitosan and the O= group of the quinone, which undergoes a Michael-type adduct as illustrated in Fig. (**8**). The high flavonoid glycoside content of this dye fraction makes it a good source of quinone molecules, compared to the original tea extracts, and thus an efficient dye source.

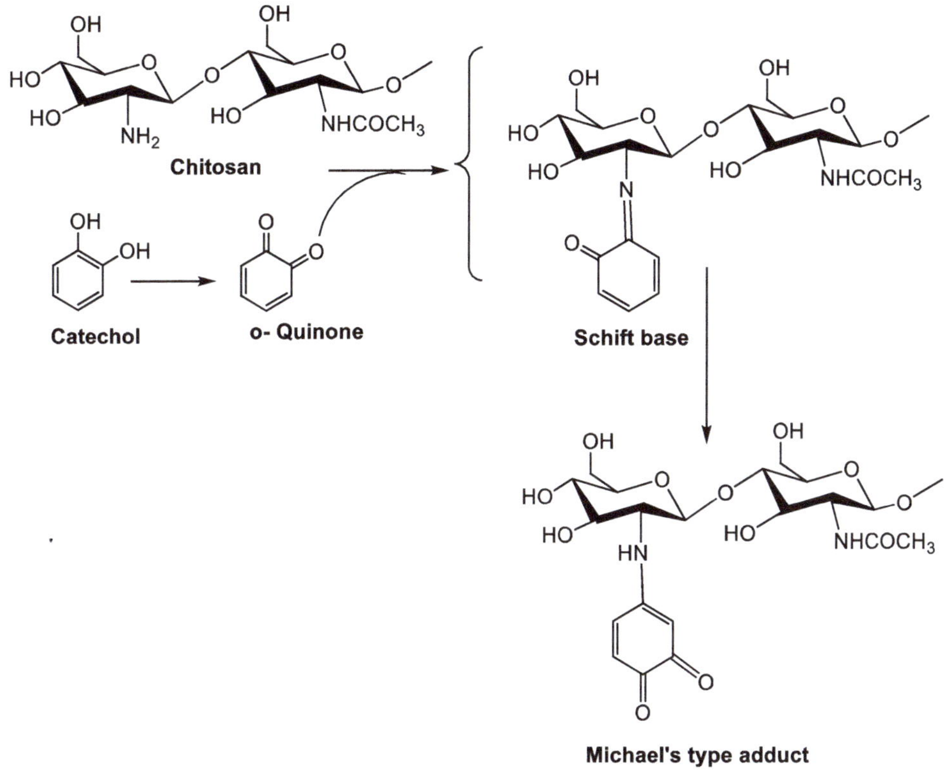

Fig. (8). Possible Chitosan–Quinone interaction.

Effect of Ligand on Dye Uptake

Cotton fabric was treated with 5 g/100 g of Tea dye in the presence of chitosan (1 g per 100 g of fabric) at different temperatures (60-80°C), and dye uptake was measured colorimetrically. As seen from Fig. (9), the dye uptake in the presence of the ligand was 2-3 times higher than that in the absence of the ligand.

	Dye Uptake%		
	60 °C	70 °C	80 °C
Dye 5 g/100 g	16.42	18.31	20.34
Dye+Chitosan (1%)	41.59	48.63	47.86

Fig. (9). Dye uptake of cotton fabric in the presence and absence of chitosan at different temperatures.

The enhanced dyeability of the modified fabric is likely a result of the reduction of coulombic repulsion between the fabric surface and the anionic dye molecules in the presence of positively charged chitosan molecules on the surface.

Tea Dye Uptake in the Presence of Metallic Mordants

When fabric (cotton, linen, nylon, or wool) was dyed with Tea and ferrous sulfate (a source of iron), it produced gray shades. The intensity of the gray varied as per the amount of tea dye used, the amount of metallic mordant used, and the type of fiber being dyed.

Fig. (**10**) illustrates the colour of cotton fabric dyed with 5g/100 g of tea dye at 70 °C, in the presence of chitosan (1 g/100 g) and Ferric oxide (1%) at 70 °C. The possible interactions between the fabric, dye molecules, and metallic mordants are also provided.

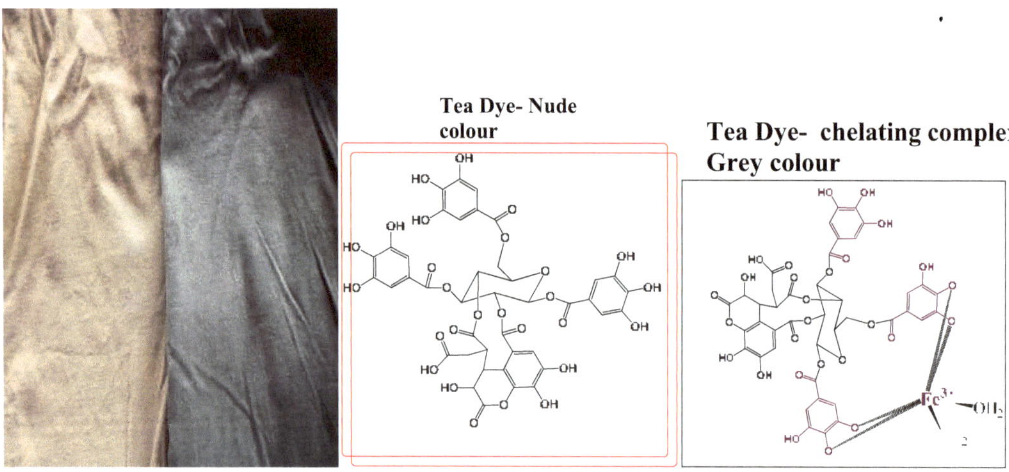

Fig. (10). Comparison of cotton fabric dyed with chitosan and metallic mordant with possible structural interactions between fabric, dye, and metallic mordants.

The gray or grayish-brown color is due to the chelating compounds produced by the interaction between phenolic compounds, such as tannins, and iron.

METHODS OF OBTAINING DIFFERENT SHADES OF TEA DYE

The most significant limitation of using this tea dye on a commercial scale was that it was only able to produce a nude colour. That is, by varying the tea dye concentration, we were only able to obtain different shades of nude colour.

To overcome this issue, the technology was further developed to obtain a full palette of colours by adding minute quantities of synthetic dyes to natural dye, with a synthetic dye to natural dye ratio of 4:96.

Dyeing Procedure to Obtain Different Shades of Fabric Using the Combination of Tea Dye and Synthetic Dye

To obtain different shades, the first step was to dye the pretreated fabrics (cotton, linen, nylon, and wool) with synthetic dyes (reactive or acidic dyes). The combination of synthetic dyes was used to obtain the required colour. However, the percentage of synthetic dyes used was below 0/5% for 100 grams of fabric. During synthetic dyeing, both Sodium sulfate and soda ash (sodium carbonate) were required. Sodium sulfate helps control the rate of dye absorption by the fibers, ensuring even color distribution. At the same time, soda ash acts as a fixative, promoting the chemical bonding of the dye to the fabric, particularly with reactive dye. The synthetic dyed fabric was washed thoroughly and treated with chitosan (0.5–1%) at an acidic pH of 4, with a temperature between 60° and 80° C for 15-30 min. The mordanted fabric were then treated with the natural dye solution, concentration ranging from 2-10% as for the fabric used (cotton, linen, nylon wool), for sufficient time (40-45 min.) at the desired temperature (70- 80 °C for cotton and linen, and 90-95 °C for nylon and wool) to produce different shades.

The colour obtained with the synthetic dye alone and after treatment with tea dye is illustrated in the following Fig. (**11**).

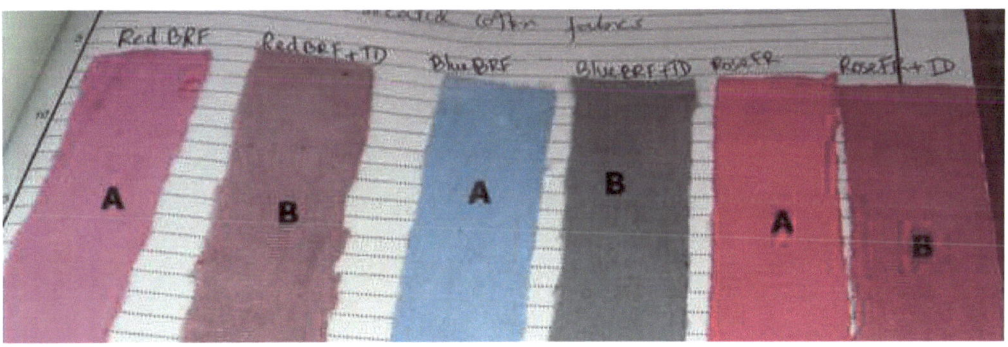

Fig. (11). Shades obtained with reactive dyes (A) and reactive dyes and tea dyes (B).

FACTORY TRIALS

After a successful laboratory illustration, bulk factory trials were conducted to scale up the process and validate the results in a larger and more practical setting.

This involved replicating the procedures and conditions from the lab on a much larger scale, utilizing specialized equipment and larger quantities of materials.

The first step in factory trials is sample trials, which were conducted using garment samples ranging from 1 kg to 15 kg. The goal of these trials was to confirm the feasibility and efficiency of the process before proceeding to full-scale production, which is termed as bulk trials.

100% Cotton garments, weighing 5 Kg (27 T-shirts), were loaded into a front-load sample dyeing machine (Vertostar Series RVW (V)) and loaded with 200 L of water. The machine is fully automatic, with the timing, water level, and chemical dosage being set through the programmer. The machine contains an outer drum and an inner drum. The outer drum holds the water and dye solution, and the inner drum is where the garments are placed and where the dyeing process occurs. The inner drum rotates both forward and backward to ensure even dyeing. The outer drum is designed to ensure minimal heat loss during the dyeing process and maintain a constant temperature for better quality and consistency.

The first step in the sample trial was pre-treatment of the garments, a crucial step for achieving uniform and high-quality dyed results. This involves treating garments with the chemical auxiliaries to remove impurities, smooth the fabric surface, improve dye absorption, and enhance color uptake. The following Table **2** outlines the steps of garment dyeing, including the added auxiliaries and their quantities, Run time, pH, Rotation Per Minute (RPM), Temperature, and Setup Time. The Liquor level was programmed for 200 L to maintain a Material to Liquid Ration (MLR) of 1:40.

Table 2. Sample Dyeing program of front-load dyeing machine for cotton garments.

S. No.	Step	Auxiliaries	Quantity	Run time/pH/ RPM/Temp/Set up Time
1	Pre-treatment	Tripon AC 20	200 g	10 /5.5/12 /70 /20
2	Rinse	Water	200 L	2/5.5/12/30/5
3	Enzyme	Biopolish CR Plus	200 g	20/4.5/12/50/10
		Acetic Acid 99.9%	20 ml	
4	Rinse	Water	200 L	2/5.5/15/30/5
5	Dyeing Auxiliaries	Lubatex ECS	200 g	5/6/15/60/20
		Sodium Sulphate	8 Kg	
6	Synthetic Dyeing	Yellow LF-R	20.65 g	20/6/15/60/0
		Red LF -3B	8.83 g	
		Royal Blue LF-3R	37.62 g	

(Table 2) cont.....

S. No.	Step	Auxiliaries	Quantity	Run time/pH/ RPM/Temp/Set up Time
7	Soda dosing and Running	Soda Ash	1.6 Kg	40/12/15/60/0
9	Water Rinse x Twice	Water	200 L x Twice	2/9/18/30/5
10	Mordant	Quaternary amine textile auxiliaries/chitosan	150 g	20/8/15/50/0
11	Tea dyeing	Acetic Acid 99.9%	40 mL	40/5/15/70/0
		Tea Dye	300 g	
9	Water Rinse x Twice	Water	200 L x Twice	5/5/12/30/5
-	-	-	-	-

Then, the garments were extracted in the Extractor Machine for 10 minutes, dried in a Hot dryer at 70°C for 100 minutes, followed by cool drying at 30°C for 20 minutes.

Many appealing shades were produced by adding different synthetic dyes to the combination of tea dye. Fig. (**12**) shows some of the colours of garments obtained in a combination of synthetic and Tea dye process in sample trials.

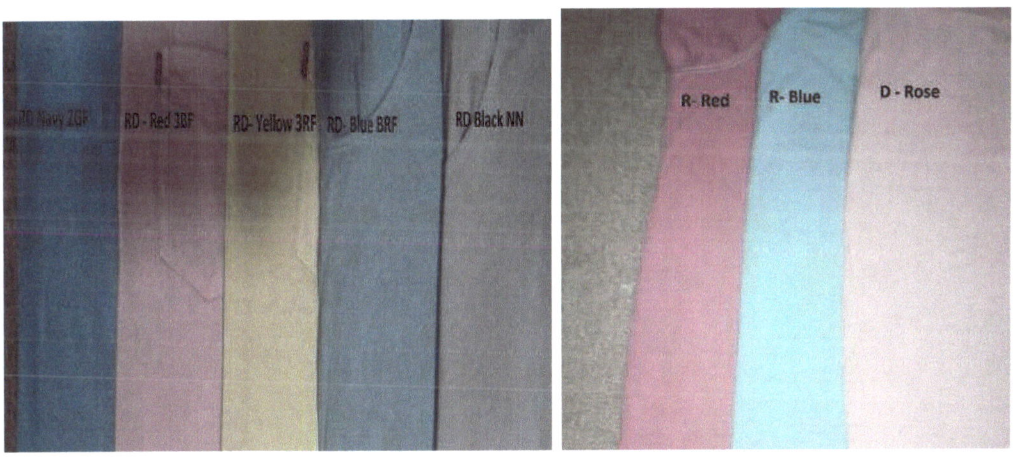

Fig. (12). Different shades of garments obtained with a combination of synthetic and Tea dye from the Sample dyeing machines for cotton (left) and nylon (right) garments.

FASTNESS PROPERTIES

American Association of Textile Chemists and Colorists (AATCC) standard protocols were used to analyze the dyed fabric for colour fastness to Light, water, perspiration, laundering, crocking, and bleaching [24].

Colour Fastness to Water, Laundering, and Crocking

Color Fastness to Water

The color fastness of dyed garments to water was measured using standard ISO-105-EO1 protocols. A specimen (4 x 10 cm) of the dyed garments are kept in contact with multifibre adjacent fabric (wool—WO, acrylic—PAC, polyester—PES, polyamide—PA, cotton—CO, or cellulose acetate—AC), and immersed in water at 37 ± 2 °C for 30 min, drained and placed between two plates under a specified pressure in a test device. The dyed garment specimen and the adjacent fabric(s) are dried separately. The change in colour of the Tea dyed specimen and the staining of the adjacent fabric(s) were assessed by comparison with the grey scales. (values 1–5, where 1—poor, 2—fair, 3—good, 4—very good, and 5—excellent). The dyed samples were folded into two layers to make them opaque while evaluating [24]. The results for the color fastness to water of tea dye fabric and tea + synthetic dyed fabric are illustrated in Figs. (**13** and **14**), respectively.

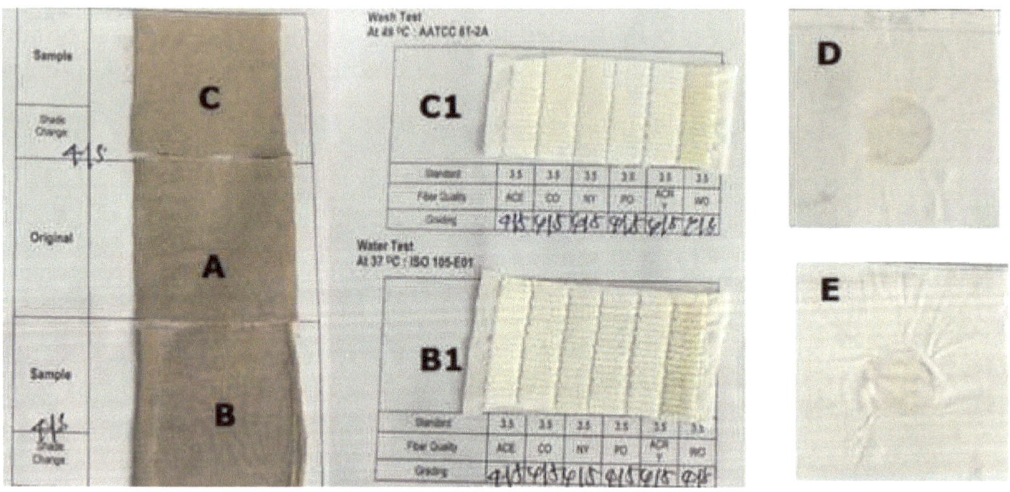

Fig. (13). Photographs of different samples in the fastness to water, wash, and rubbing test; (**A**) original Tea dyed sample, (**B**) specimen after water fastness test, and **B1** is its adjacent fabric, (**C**) specimen after wash fastness test, and **C1** is its adjacent fabric. **D** and **E** are specimens of rubbed test fabrics, **D** for wet rubbing and **E** for dry rubbing.

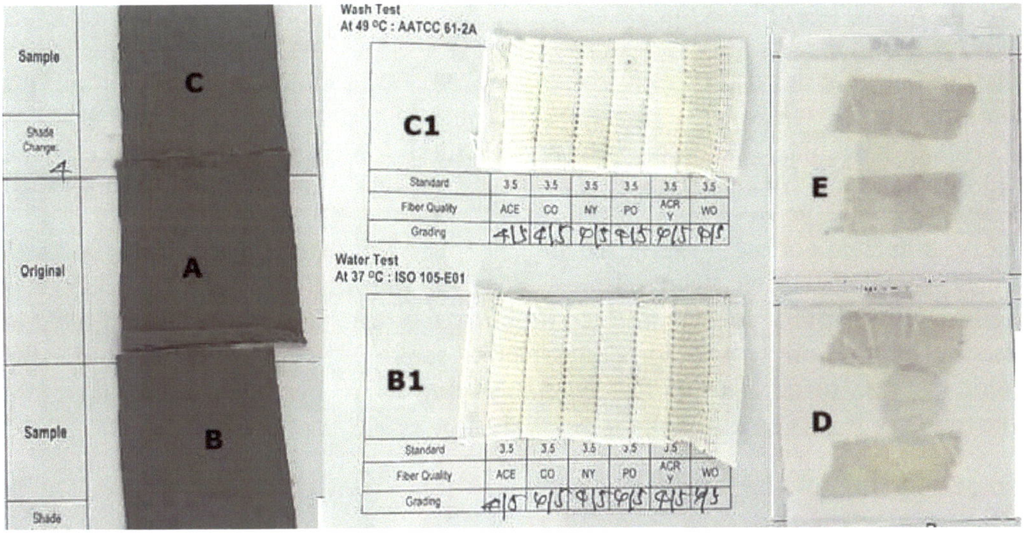

Fig. (14). Photographs of different samples in the fastness to water, wash, and rubbing test; (**A**) original Tea + Synthetic dyed sample, (**B**) specimen after water fastness test, and **B1** is its adjacent fabric, (**C**) specimen after wash fastness test, and **C1** is its adjacent fabric. **D** and **E** are specimens of rubbed test fabrics, **D** for wet rubbing and **E** for dry rubbing.

Colour Fastness to Laundering

The color fastness to laundering of dyed garments was carried out using an accelerated laundering machine, as per AATCC 61, the standard test method. This test method simulates the effects of multiple home launderings on fabrics to assess their colorfastness and potential for staining. For the washing test, a 50 × 150 mm sample fabric was attached to a piece of multifiber-adjacent fabric (wool—WO, acrylic—PAC, polyester—PES, polyamide—PA, cotton—CO, or cellulose acetate—AC). They were washed separately in stainless steel canisters, with the liquor volume in each canister set to 150 mL, and 50 steel balls were added to each. The non-ionic PERLAVIN NIC detergent (0.15%) was used as the reference detergent. The washing water temperature was set to 49 °C, while the washing duration was 45 min. After a single washing test, the samples were rinsed three times with distilled, hard, and soft water and were then air-dried. The color loss of specimens and the staining of adjacent fabrics were graded by comparison with the grey scales (values 1–5, where 1 = poor, 2 = fair, 3 = good, 4 = very good, and 5 = excellent) [24]. The results for the color fastness to laundering of tea dye fabric and tea + synthetic dyed fabric are illustrated in Figs. (**13** and **14**), respectively.

Colour Fastness to Crocking

The amount of color transferred from a colored textile material to another surface through rubbing was measured using a rubbing fastness tester, namely the James H. Heal Crockmeter. The standard test method AATCC 8, also known as the Colorfastness to Crocking Test, was implied.

A dry, bleached cotton cloth measuring 50 × 130 mm was rubbed back and forth on the dyed specimen twenty times. For wet rubbing, the rubbing cloth was wetted with distilled water before conducting the test. The greyscale was used to rate the staining on the rubbing cloths (rating range 1–5) [24]. The results for the color fastness to rubbing of tea dye fabric and tea + synthetic dyed fabric are illustrated in Figs. (**13** and **14**), respectively.

Colour Fastness to Light

To measure the tea-dyed fabric's colour fastness to Light, the standard method AATCC 16.3 was used. This method assesses how well the tea-dyed material retains its color when exposed to Light. The test involves exposing the fabric to a Xenon arc lamp and then comparing the exposed sample to an unexposed control sample to evaluate color change. The test samples, measuring 70 x 120 mm in size, are exposed to a Xenon arc lamp that emits radiation from below 270 nm in the UV spectrum to the visible spectrum and into the infrared, simulating the effects of sunlight. The measurement temperature was 47 °C, while the relative humidity was 40%, and the exposure time was 40 hours. The color change after exposure is compared with the grey scale from 1 = worst to 5 = best [24].

Colour Fastness to Non-Cl_2 Bleach and Perspiration

Colour fastness to non-Cl_2 bleach was tested using AATCC/ASTM TS-001, and color fastness to perspiration using AATCC 15-13. The AATCC/ASTM TS-001 test method is designed to evaluate the colorfastness of textiles to powdered non-chlorine bleach in home laundering. The procedure involves applying the bleach solution, allowing it to penetrate the fabric under appropriate temperature conditions, and then subjecting it to the abrasive action of five home laundering cycles, followed by rinsing and evaluation for color change.

The AATCC 15-13 assesses how well a tea-dyed textile retains its color when exposed to simulated perspiration. The test utilizes a chemical solution that mimics the composition of human perspiration, encompassing both acidic and alkaline components. The test involves wetting the textile specimen with a perspiration solution, placing it in a perspirometer (a device used to apply pressure), and then compressing it. The compressed specimen is heated (38°C) for

6 hours. After drying, the specimen is assessed for color change using the Gray Scale, with a higher grade indicating better colorfastness [24].

Fastness properties with respect to Light, water, perspiration, laundering, crocking, and bleaching for the factory sample trials with respect to cotton and nylon fabric for both only tea dye and various shades of tea dye obtained through the addition of relevant synthetic dyes are tabulated in Table **3**.

Table 3. Colours of Hybrid dyes and their Fastness properties with respect to Light, water, perspiration, laundering, crocking, and bleaching.

Synthetic dye	Cotton colour	Colour fastness to							
		Non-Cl_2 bleach	Light	Water	Pers-piration	Laundering		Crocking	
						Actual	Accelerated	Dry	Wet
Only tea dye		4.0	3.5	4.5	4.5	3.5	4.0	4.5	3.5
Chrofix Yellow 3RF		4.5	3.5	4.5	4.5	3.5	4.0	4.5	3.5
Chrofix Orange CD		4.5	3.5	4.5	4.5	3.5	4.0	4.5	3.5
Chrofix Red 3BF		4.5	3.5	4.5	4.5	4.0	4.0	4.5	3.5
Chrofix Blue BRF		4.5	3.5	4.5	4.5	3.5	4.0	4.5	4.5
Chrofix Blue RSP		4.5	3.5	4.5	4.5	3.5	4.0	4.5	3.5
Chrofix Turq Blue G		4.0	3.5	4.5	4.5	3.5	4.0	4.5	3.5
Chrofix Navy 2GF		4.5	3.5	4.5	4.5	4.0	4.0	4.5	3.5
Novacron Black N (Huntsman)		4.5	3.5	4.5	4.5	3.5	4.0	4.5	4.0
	Nylon	-	-	-	-	-	-	-	-
Only tea dye		4.5	4	4.5	4.5	4.5	4.0	4.0	3.5
Sellan Red GB		4.5	4.5	4.5	4.5	4.5	4.0	4.0	3.0
Tellon Navy AMF		4.5	4	4.5	4.5	4.5	4.0	3.5	3.0

(Table 3) cont.....

Synthetic dye	Cotton colour	Colour fastness to							
		Non-Cl_2 bleach	Light	Water	Pers-piration	Laundering		Crocking	
						Actual	Accelerated	Dry	Wet
Sellan Blue GP		4.5	4.5	4.5	4.5	4.5	4.0	4.0	3.5
Ciblan Black MSRL		4.5	4.5	4.5	4.5	4.5	4.0	4.0	3.5

ANTIBACTERIAL ACTIVITY OF TEA-DYED GARMENTS

The assessment of antibacterial activity of tea dyed garments was measured using AAATCC TM 100-2012.

During the procedure, tea-dyed swatches and control swatches (100% cotton fabric) were treated with a bacterial inoculum (1.0 ± 0.1 mL). The test organisms used in the assay were *Staphylococcus aureus*, American Type Culture Collection (ATCC) number 6538, and *Klebsiella pneumoniae*, ATCC number 4352. The concentration of *S. aureus* was 1.5×10^{5} colony-forming units (cfu)/mL, and for *K. pneumoniae* it was 1.7×10^{5} cfu/mL. The bacterial-inoculated samples were incubated at 37 °C for 24 hours. After incubation, the swatches are thoroughly mixed in a vortex with a neutralizing broth solution to extract all the bacteria into the solution. The neutralizing solution, which contains the bacteria from the fabric, is sequentially diluted, and a small amount of it is inoculated in an agar plate, and the number of colonies that grow on the plate is counted [25].

The results indicated that the tea-dyed fabric was able to reduce the bacterial growth of *S. aureus* by 99% and a 98% reduction in the case of *K. pneumoniae*. To further evaluate the effect of washing on the antibacterial activity of tea-dyed samples, the tea-dyed swatches were subjected to 15 washes and then subjected to the above assay. The results indicated a 98% reduction in the case of *S. aureus* and a 96% reduction in the case of *K. pneumoniae*, demonstrating that the tea-dyed fabric retained its antibacterial activity fully even after 15 washes.

LIFE CYCLE ASSESSMENT OF TEA DYE *VS.* SYNTHETIC DYES

A Life Cycle Assessment (LCA), a tool that enables the evaluation of a product or product system's potential environmental impact, was conducted to assess the ecological impact of the tea dye powder compared to synthetic dyes. The study calculated the total greenhouse gas emission per functional unit (dyeing one kilogram of fabric) in terms of CO_2 equivalent using the IPCC 2013 GWP 100a approach [26]. The study revealed that acid dyeing, disperse dyeing, and reactive dyeing were responsible for 726 g, 737 g, and 935 g of CO_2 equivalent emissions, respectively. In contrast to the synthetic dyeing process, natural dyeing offers a 393 g CO_2 equivalent saving (Fig. **15**).

An LCA study was also carried out to compare the dyeing process of cotton garments using 100% synthetic dye and Natural tea dye, in the following ratios: 50:50 (I), 60:40 (II), 70:30 (III), 80:20 (IV), and 90:10 (V). According to the results obtained during the dyeing process of 1 kg of cotton, with the above-mentioned composition, the CO_2 equivalent emissions were 1.055 Kg, 0.495 Kg, 0.459 Kg, 0.433 Kg, 0.411 Kg, and 0.396 Kg, respectively. Thus, compared with synthetic dyeing, Natural + synthetic dyeing of types I, II, III, IV, and V shows 53.1%, 56.5%, 59.0%, 61.0%, and 62.5% decrease in CO_2 equivalent emission, respectively.

SUSTAINABILITY AND COMMERCIAL APPLICATION OF WASTE TEA DYE

Sustainability is a hot topic in the textile industry. A textile dyeing factory would be very much interested in utilizing natural resources as a textile dye. Still, it would not be feasible for the textile factory to produce the natural resources on its own. Therefore, a scientific solution should be such that it is attainable for application in industrial-scale approaches, enabling the development of sustainable and environmentally friendly textile goods in sufficient quantities to meet the needs of the textile sector.

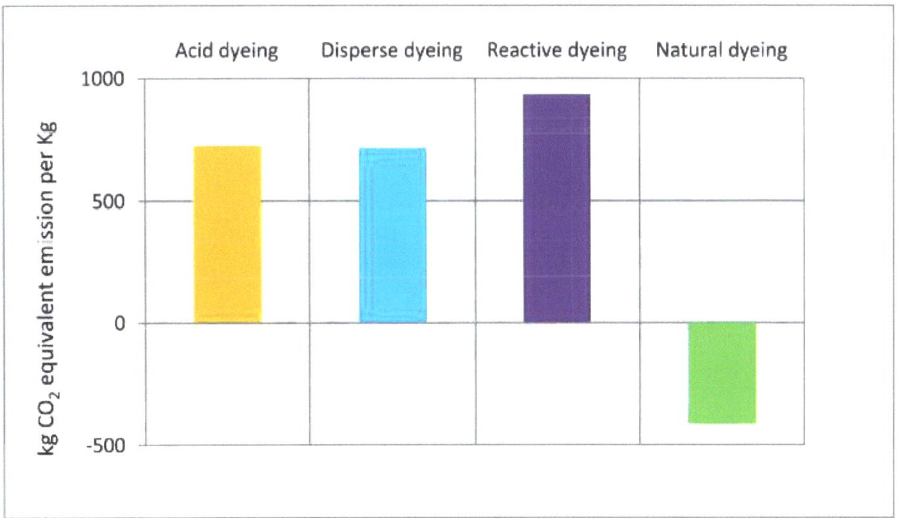

Fig. (15). CO_2 footprint for reactive, acid, disperse, and tea dye.

The idea of using waste from one factory as a raw material for another is a win-win scenario, promoting resource efficiency and circular economy principles.

The Ready-to-Drink manufacturing plant was discarding ¼ of hot water-soluble tea extracts, mainly consisting of tannins and other conjugated phenolic compounds, to make its Ready-to-Drink Tea soluble in cold water. The discarded fraction from one factory (the largest RTD manufacturing plant) accounted for about 700-1000 MT annually, which accounts for an enormous cost for its disposal. The textile industry is increasingly adopting greener technologies due to rising consumer demand for sustainable practices and the implementation of stricter environmental regulations. On the other hand, consumers are demanding products that come from responsible sources, but at the same time, they are looking for lower costs. To facilitate the commercial application of tea dye, all the above challenges were taken into consideration. By demonstrating how RTD waste can be utilized as a practical and sustainable source of natural dye, this work addressed the needs of both industries and the environment. As no additional materials, equipment, or advanced technologies were required, it was soon adopted and scaled up. The method was exaggerated to obtain different shades utilizing tea dye along with minute quantities of synthetic dyes. The technology has been adopted by numerous reputable textile brands worldwide.

CONCLUSION

The current study demonstrates the practical application of waste tea as a sustainable and scalable natural dye source, addressing both environmental concerns and industrial requirements.

The study demonstrated how a bio-waste product, specifically tea cream generated during the Ready-To-Drink (RTD) instant tea powder manufacturing process, can be converted into an effective dye using simple techniques. Chemical analysis of tea cream using LCQTOF MS/MS established it to be a rich source of polyphenolic compounds, with differences in composition as compared to the original hot water extract of Broken Mixed Fannings (BMF). The chemical analysis revealed that the tea dye was rich in flavones/isoflavones glycosides, which provided the required color as well as the hydrophobicity necessary for enhancing dye uptake to the hydrophobic fibre. Furthermore, the resistance to washing was achieved by the addition of polyamines, such as chitosan, which enabled the formation of a bond between the fabric fibre and the tea dye, resulting in the required fastness properties.

Tea dye was effective as a colourant for cotton, linen, nylon, wool, and their blends, with effective tea dye concentrations varying from 1 to 10% depending on the type of fabric and the required colour. The technology was further enhanced to obtain a full palette of colours by adding minute quantities of synthetic dyes to natural dyes, with a synthetic dye to natural dye ratio of 5:95. The lab-scale results were upscaled to sample scale and then finally to bulk scale in the textile industry. The fastness properties of the dyed fabric were tested in accordance with the industrial standards. The results revealed that the tea-dyed fabric met the required standards for fastness properties, including washing, laundering, rubbing, perspiration, bleaching, and light fastness. Additionally, as per the AAATCC TM 100-2012 protocols, the tea dyed fabric exhibited antibacterial activity against *S. aureus* and *K. pneumoniae*. Studies also revealed that the antibacterial effect of tea-dyed fabric was retained even after 15 washes. The study presents a readily applicable innovative green solution for coloring fabrics with tea waste generated during the Ready to Drink tea powder.

In future studies, the technology will be advanced to utilize the tea dye for dyeing polyester fabrics.

As biowaste was utilized as a fabric colourant, it results in negative CO_2 emissions compared to synthetic dyes.

The idea of using waste from one factory as a raw material for another is a win-win scenario, promoting resource efficiency and circular economy principles. The findings of this research have been implemented by various reputed brands worldwide.

CONFLICT OF INTEREST

None of the authors has a specific conflict of interest with the focus of this study. Vinitha M. Thadhani has been an employee of the Sri Lankan Institute of Nanotechnology. (July 2014- June 2020). V. M. Thadhani has not received any fees or honoraria since June 2020. This work was initiated during her previous position at the Sri Lankan Institute of Nanotechnology in affiliation with the H.E.J. Research Institute, University of Karachi, Pakistan. The authors declare no competing financial interest related to the content presented in this chapter

ACKNOWLEDGMENTS

The authors wish to recognize the application of AI tools, like Grammarly, during the writing process to enhance the clarity and language of the manuscript. The authors thoroughly reviewed and modified the content as necessary and accept full responsibility for the content of the chapters. Additionally, they want to emphasize that AI was not utilized in their chapters to create or modify images, aid in data analysis, or contribute to their research.

REFERENCES

[1] Lellis B, Fávaro-Polonio CZ, Pamphile JA, Polonio JC. Effects of textile dyes on health and the environment and bioremediation potential of living organisms. Biotechnology Research and Innovation 2019; 3(2): 275-90. [http://dx.doi.org/10.1016/j.biori.2019.09.001]

[2] Jabeen N, Jabeen A. To identify hazards and risks (Health and safety) in the textile dyeing industry. Asian J Home Sci 2017; 12(1): 182-7. [http://dx.doi.org/10.15740/HAS/AJHS/12.1/182-187]

[3] Bailey K, Basu A, Sharma S. The Environmental Impacts of Fast Fashion on Water Quality: A Systematic Review. Water 2022; 14(7): 1073. [http://dx.doi.org/10.3390/w14071073]

[4] Al-Tohamy R, Ali SS, Li F, *et al.* A critical review on the treatment of dye-containing wastewater: Ecotoxicological and health concerns of textile dyes and possible remediation approaches for environmental safety. Ecotoxicol Environ Saf 2022; 231: 113160. [http://dx.doi.org/10.1016/j.ecoenv.2021.113160] [PMID: 35026583]

[5] Synthetic Dye Global Market Report 2024, The Business Research Company, Published Jan 2024. Available from: https://www.thebusinessresearchcompany.com/report/synthetic-dye-global-mar-et-report

[6] Har Bhajan Singh H. B and Kumar, A. B. Handbook of natural dyes and pigments Handbook of natural dyes and pigments. Woodhead Publishing India Pvt. Ltd. New Delhi, 2014.

[7] Vankar PS, Shukla D, Wijayapala S. Low temperature optimized dyeing of cotton, wool and silk with extract of *Camellia sinensis* (tea leaves). Journal of Textile Engineering & Fashion Technology 2017; 2(1): 274-80. [http://dx.doi.org/10.15406/jteft.2017.02.00045]

[8] Hossain T, Das SC, Akhtarujjaman M, Uddin MA, Bedoura S. Eco-friendly dyeing of cotton fabric using used tea bag extracts: A comprehensive study on pH, fabric cationisation, and computational analysis. Current Research in Green and Sustainable Chemistry 2023; 7: 100387. [http://dx.doi.org/10.1016/j.crgsc.2023.100387]

[9] Adeel S, Ahmad M, FazalurRehman, *et al.* Eco-friendly dyeing of cotton fabric with waste tea leaves-based tannin natural dye. Glob NEST J 2021; 23(3): 365-9.

[10] Ordu DS, Ömeroğulları Z. Sustainable textiles colored with tea waste in the printing process. Tekstil ve Mühendis 2025; 32(137): 44–55. [http://dx.doi.org/10.7216/teksmuh.1593823]

[11] Sarker MKU, Haque MM, Hasan MR, Sultana S, Ray SK, Shaikh MAA. Utilization of factory tea (*Camellia sinensis*) wastes in eco-friendly dyeing of jute packaging fabrics. Heliyon 2024; 10(10): e30948. [http://dx.doi.org/10.1016/j.heliyon.2024.e30948] [PMID: 38779026]

[12] Hassan SRB, Abdul Rahim NABC. Extraction of natural dye from black tea waste for cotton dyeing application. Journal of Engineering Science and Technology 2021; 16(3): 2562-70.

[13] Sukemi , Pratumyot K, Srisuwannaket C, Niamnont N, Mingvanish W. Dyeing of cotton with the natural dye extracted from waste leaves of green tea (*Camellia sinensis* var. *assamica*). Color Technol 2019; 135(2): 121-6. [http://dx.doi.org/10.1111/cote.12381]

[14] Triwiswara M, Indrayani L. Utilization of black tea waste as natural batik dyes on cotton and silk. IOP Conf Ser Earth Environ Sci 2020; 456(1): 012051. [http://dx.doi.org/10.1088/1755-1315/456/1/012051]

[15] Preetha R, Rani NA. Dyeing at home from kitchen waste - tea and coffee residue. ShodhKosh J Vis Per Arts 2023; 4(2): 374-84.

[16] Xu H, Song H. A Systematic Review of Tea Dyeing in the Textile Industry: Research Trends, Dyeing Performance, Functional Characteristics, and Application Potential. AATCC Journal of Research 2025; 12(1): 24723444251314916.
[http://dx.doi.org/10.1177/24723444251314916]

[17] Houshyar S, Amirshahi SH. Treatment of cotton with chitosan and its effect on dyeability with reactive dyes. Iran Polym J 2002; 11: 295-302.

[18] Iced Tea - Market Share Analysis, Industry Trends & Statistics, Growth Forecasts 2024 – 2029. Mordor Intelligence. (2024). Available from: https://www.giiresearch.com/report/moi1408145-ice--tea-market-share-analysis-industry-trends.html

[19] Dalpathadu KAP, Rajapakse HUKDZ, Jayasinghe CVL, Nissanka SP. Manufacturing of Low Haze Instant Tea Extracts Using Sri Lankan Broken Mixed Fannings (BMF) as Raw Materials. Tropical Agricultural Research 2022; 33(4): 319-29.
[http://dx.doi.org/10.4038/tar.v33i4.8585]

[20] Thadhani V.M, Amaratunga, Gehan, Akhtar Naheed, Musharraf Syed Ghulam, Choudhary Muhammad Iqbal, Method of extraction of an effective tea dye powder from teas waste and application thereof on fabric and garments. (2021) WO2021033173.

[21] Rabbi MBU, Al Montasir MS, Emon EH, *et al.* Enhancing the Antimicrobial and UV Resistance Properties of PC Blend and Recycled Cotton Fabric through Ag/TiO2 Nanocomposite. ACS Omega. 2025; 10(8): 7697-710.
[http://dx.doi.org/10.1021/acsomega.4c07348] [PMID: 40060846] [PMCID: PMC11886643]

[22] Hayat T, Adeel S, Fazal-ur-Rehman , *et al.* Waste black tea leaves (Camelia sinensis) as a sustainable source of tannin natural colorant for bio-treated silk dyeing. Environ Sci Pollut Res Int 2022; 29(16): 24035-48.
[http://dx.doi.org/10.1007/s11356-021-17341-5] [PMID: 34822080]

[23] Akhtar N, Thadhani VM, Ul Haq F, Khan MN, Ali S, Musharraf SG. Rapid identification and quantification of bioactive metabolites in processed *Camellia sinensis* samples by UHPLC-ES--MS/MS and evaluation of their antioxidant activity. J Ind Eng Chem 2020; 90(25): 419-26.
[http://dx.doi.org/10.1016/j.jiec.2020.07.047]

[24] Manual of International test methods and procedures (2022). The American Association of Textile Chemists and Colorists, 2022. Volume 97.

[25] Assessment of Antibacterial Finishes on Textile Materials. American Association of Textile Chemists and Colorists (AATCC). AATCC TM 100: Certified Reference Materials for Antimicrobial Testing.

[26] Environmental Management-Life Cycle Assessment-Principles and Framework, International Standard ISO14040, 2006. International Organization for Standardization: CH-1211 Geneve 20 l Switzerland.

SUBJECT INDEX

D

E

F

G

H

I

J

K

L

M

N

O

P

Q

R

S

T

U

V

W

www.ingramcontent.com/pod-product-compliance
Lightning Source LLC
LaVergne TN
LVHW070951210926
848482LV00002B/181

* 9 7 9 8 8 9 8 8 1 5 4 2 4 *